Photos by Ingbet Gruttner

ABOUT THE AUTHORS

SCOTT BITTLE is executive editor of Public Agenda.org, where he has prepared citizen guides on more than twenty major issues including the federal budget deficit, Social Security, and the economy. He is also the website director for *Planet Forward*, an innovative PBS program designed to bring citizen voices to the energy debate. A former newspaper reporter and bureau chief, he won two Golden Quill Awards for feature articles and was honored by the Philadelphia Press Association.

A cofounder of Public Agenda.org, JEAN JOHNSON has written articles and op-eds for *USA Today*, *Education Week*, *School Board News*, *Educational Leadership*, and the *Huffington Post*. She is an advisor to the Ad Council and on the board of the National Issues Forums, a national nonpartisan citizen engagement coalition.

WHO
TURNED OUT
THE LIGHTS?

For Susan

Also by Scott Bittle and Jean Johnson

Where Does the Money Go?

WHO TURNED OUT THE LIGHTS?

Your Guided Tour to the Energy Crisis

SCOTT BITTLE & JEAN JOHNSON

HARPER

NEW YORK · LONDON · TORONTO · SYDNEY

HARPER

HarperCollins books may be purchased for educational, business, or sales promotional use. For information please write: Special Markets Department, HarperCollins Publishers, 10 East 53rd Street, New York, NY 10022.

FIRST EDITION

Designed by Betty Lew

Library of Congress Cataloging-in-Publication Data

Bittle, Scott.
    Who turned out the lights? : your guided tour to the energy crisis / Scott Bittle and Jean Johnson. — 1st ed.
        p. cm.
    Includes bibliographical references.
    ISBN 978-0-06-171564-8
    1. Power resources—Popular works. 2. Energy conservation—Popular works. 3. Energy consumption—Popular works. I. Johnson, Jean. II. Title.
    TJ163.2.B578 2009
    333.79—dc22

                                                                2009018884

---

09   10   11   12   13   WBC/RRD   10 9 8 7 6 5 4 3 2 1

Everybody's talkin' at me.

I don't hear a word they're saying . . .

—*Fred Neil, soundtrack to* Midnight Cowboy

CONTENTS

||||||||||||||||||||||||||||||||||||||||||||||||||||||||||||||||||||||||||||||

PREFACE

WHERE WE'RE COMING FROM

In the last couple of years, just about everyone who's anyone has gone on TV to tell Americans what the United
States should do to solve its energy problems: John Mc-
Cain, Barack Obama, Sarah Palin, Al Gore, T. Boone Pickens, Sheryl Crow. In the midst of the 2008 presidential
campaign, none other than Paris Hilton paused poolside to
share her views on energy, reassuring us that she is "like
totally ready to lead" should the need arise. (Good to know,
Paris. We'll get back to you on that one.)

Yet despite being inundated with information about the
country's energy troubles, many Americans admit that
they're pretty confused.[1] It's a perfectly understandable
condition. In the past four decades, the country has taken
the same approach to energy that Homer Simpson takes to,
well, everything. There's near panic when energy prices go
up (d'oh!) and complacency when they head downward
(woo-hoo!). Experts spar with each other over hugely important questions such as whether the world will start running out of oil in our lifetimes and whether global warming

will cause crop failure and famine. But the debate is just as ferocious on matters that are closer to home: does eating local produce and using cloth diapers really save energy or not, and is it enough to matter?

For many of us, the nation's energy debate has become an incomprehensible jumble, so the purpose of *Who Turned Out the Lights?* is to stop, take a deep breath, back up a bit, and go back to basics. We've written this book because we're convinced that there are millions of Americans who are concerned about the energy issue and want to understand it better. Our aim is to explain the nuts and bolts in plain, solid, nonscientific, nontechnical English. We also believe that increasing numbers of Americans recognize that the country simply has to stop arguing about energy and start doing something about it. Our plan is to describe the chief options as understandably as we can and summarize different points of view about them. We're not recommending solutions here. Instead, we're trying to offer enough perspective, context, and information so readers can stop relying on Hollywood stars and pundits for direction. We want to help you decide for yourself what path the United States should take.

WHO WE ARE

On any controversial issue, it's always a good idea to know who is writing what you're reading and where they're coming from. We both work for a nonprofit, nonpartisan research organization called Public Agenda that covers policy issues and conducts public opinion research.* In our jobs, we have good access to people who are knowledgeable about issues such as energy and the environment. We have

---

* You can find out more at www.publicagenda.org.

also spent years translating expert information into terms and concepts nonexperts can understand. *Who Turned Out the Lights?* is our second book explaining a seemingly impenetrable, convoluted issue for typical citizens. Our first—*Where Does the Money Go? Your Guided Tour to the Federal Budget Crisis*—looked at the U.S. government's massive debt and the unsustainable finances of Social Security and Medicare.

Here's some of the thinking behind this book:

- **First, our goal is to explain the basics.** Essentially, we see ourselves as translators. We've spent more than a year reviewing information from government, the energy industry, environmental organizations, and independent sources. Ironically, the fact that we're not energy experts ourselves has been an asset. We've been through a learning curve on this issue, and we can remember moments when we were confused as all get-out. Having been through it, we'd like to think we can help you get through it too.

- **Second, our aim is to present options, not make recommendations.** There are plenty of people who will tell you what they think the country should do, and quite a few of them have launched glossy ad campaigns to make sure you hear their ideas. There are thousands of books, articles, and op-eds recommending specifics. That's not what we're about here. Instead, we'll give you some basic facts and lay out the choices so you can come to your own conclusions. Naturally, there is plenty of expert disagreement; in those cases we try to help you understand what the experts are arguing about. When there seems to be a pretty strong consensus of expert opinion one way or the other, we tell you that. There are some areas where, truth be told, the experts don't actually know the answer. The best of them admit it, and we pass that along too.

- **Third, we believe the country is ready to act.** It's easy to get frustrated about this issue because the United States has basically been in a holding pattern for decades. The blizzard of contending facts and figures can be mind-boggling. The arguments and counterarguments can be confusing, but there is way too much at stake to keep procrastinating. If this country doesn't get its energy act together, we could pull the rug out from under our economy and leave our children living in a world that bears little resemblance to the one we know now. The good news is that the national debate has become more purposeful in the last few years. In Washington, in the states, in boardrooms, in the environmental movement, in universities, research centers, and think tanks nationwide, energy has moved to the top of the agenda. There are plenty of ideas and increasing experimentation and openness to compromise. We're optimistic that the country is beginning to get off the dime.

- **Fourth, decisions now have big implications down the road.** Almost every idea anyone has about addressing the country's energy problems will take years to kick in, and decisions we make (or don't make) now will have major repercussions down the road. Power plants are built to last decades.[2] It takes years for Detroit to retool factories and reorganize to get different cars to market. Researchers need time to develop new fuels and clean up old ones, and then the best ideas have to be put into practice and moved out onto the market. Changes in tax policies and marketplace incentives work gradually, not overnight. Almost all experts agree that we need to work on a lot of different ideas on different fronts. So the point is? No silver bullet. No quick fix. We need to get started now.

- **Fifth, this issue changes all the time.** Gas prices go up and down; world events rock the price of oil; new reports, estimates, projections, and discoveries emerge regularly. Legislation gets passed, and government policies shift. But the key facts and challenges are remarkably constant. The United States is heavily dependent on imported oil and other fossil fuels that endanger the environment. We're competing worldwide for energy that, in the future, could be in increasingly short supply. As we take this book to press, there are important moving targets to keep an eye on. In 2009, the U.S. Environmental Protection Agency ruled that "greenhouse gas pollution is a serious problem now and for future generations;"[3] regulations on using coal, oil, and natural gas may be revised as a result. Congress is considering policies to cap the emissions that contribute to global warming. The auto industry and the Obama administration have agreed on higher mileage standards for new cars. There is plenty of stuff in the works, no doubt about it. But believe us, based on everything we have learned, whatever happens in the next year or two is just the beginning. Changing the way the United States gets and uses energy will take decades. It will be the sum total of countless decisions at all levels—by the federal, state, and local governments, by businesses, and by individuals. We'll be working on this one a long, long time.

- **Sixth, we're focusing on public questions, not individual ones.** There's been plenty of advice—and not a little lecturing—about what we should all do in our everyday lives to save energy and "go green." There's information galore on what kinds of cars to buy, how to use them, how to make our living spaces more energy-efficient—turn out the lights, keep your tires properly inflated, buy Energy Star appliances, and so on. Please, go out and do everything you can to use energy wisely. If you want some specific ideas about what you can do personally, there's a list of organizations and websites offering precisely this kind of information in the online appendix at www.publicagenda.org/whoturnedoutthelights. But it won't be enough for each of us to do our personal best. We can't personally decide what kind of power plants our local utility company will build. We can't buy cars and appliances that aren't on the market. Unless we're Silicon Valley billionaires, we can't personally fund the research that will be needed to solve our energy problems or give promising start-ups the money they need to bring innovations to the market. We can't all build our own state-of-the-art, fully functional solar house. Maybe Ed Begley Jr. can. Most of us can't. There are public decisions that have to be made, and our country and communities need to start making them. See Chapter 15 for a summary of some of the major ones.

- **Seventh, we're not refighting the climate change war here—we have too much else to cover.** We're into full disclosure, so let's get this out of the way right up front. What we're writing here is based on the assumption that the country needs to reduce activities that contribute to global warming. Everything we've learned researching this book suggests that most scientists, government officials, and even leaders in the energy industry itself accept the idea that global warming is real, that human activity is the leading cause, and that, consequently, we need to limit carbon dioxide emissions. There is a range of views on exactly what the results of global warming may be and how quickly they may occur. Even so, our belief is that the country needs to start moving away from its heavy reliance on fossil fuels—for a whole host of reasons. If you'd like to review the arguments on global warming for yourself, our online appendix includes suggestions on where you can read up on both sides, but we're not going to revisit the debates here. Even if you're not 100 percent sure about global warming, consider the risk posed by the fact that more and more people worldwide want oil, and that it's almost bound to get more costly. Consider the risk posed by the fact that the United States relies on so much imported oil, and the countries with the biggest reserves aren't exactly our best friends. As far as we're concerned, here's the takeaway: The way we use energy now poses three risks to our future—potential damage to the Earth, possible harm to the U.S. economy, and having to depend on other countries—good, bad, and otherwise—for vital energy supplies. Maybe you don't care about all three, but Meat Loaf was on to something with his "two out of three ain't bad." Frankly, just one is enough of a reason to get a move on.

WILL WE DO IT THE EASY WAY OR THE HARD WAY?

The United States has prospered for a good long time in an economy and society based on using oil, coal, and natural gas—all fossil fuels. It's been a great ride in more ways than one. Nearly everything we do depends on these energy sources to some extent, so moving away from them—or changing how we use them—will be a massive shift in how we live. To use the old TV tough-guy formulation, we can do it the easy way or the hard way.[4]

The easy way would be to think through our options, make some decisions, and get on with the job of revamping how the United States uses energy. Most of the solutions will cost money and require time, effort, and adjustment from nearly all of us, but it's entirely doable if we decide to do it. There may even be some enticing business opportunities developing the new energy technologies needed to move away from fossil fuels. The hard way would be to continue with the wasteful, head-in-the-sand energy policies we've pursued in the past. Not too far down the road, decisions about our economy and our way of life could be taken out of our hands. We'll have to scramble to buy increasingly high-priced energy from people who are thinking about their own priorities, not about what's good for the United States. And we could lose more than we even know by letting the planet slide more deeply into global warming.

A HERCULEAN CHOICE

We can forgive you if you're not optimistic about the country choosing the easy way. Human beings often have a way of ignoring even the most sensible, obvious warnings. It's a little like something that happened on the amiably cheesy TV show *Hercules: The Legendary Journeys*. Hercules, played by Kevin Sorbo, gives the bad guys the choice of the easy way or the hard way. Naturally, being bad guys, they rush him anyway. "Nobody ever wants to do it the easy way," he sighs, and then he pummels them.

But the United States has reached the point on energy where we can no longer afford to behave like a thickheaded television henchman and ignore the warnings right in front of us. We've got to get smart about this, look at the options clearly and without wishful thinking, and make sensible decisions that will really stick. We can't afford to take the drubbing that goes along with choosing the hard way.

WHO
TURNED OUT
THE LIGHTS?

||||||||||||||||||||||||||||||||||||||||||||||||||||||||||||||||||||||||||||||||||||||

CHAPTER 1

SIX REASONS THE UNITED STATES NEEDS TO GET ITS ENERGY ACT TOGETHER

You can't always get what you want.
—*The Rolling Stones*

We don't know exactly what the price of gas will be by the time you have this book in your hands, but whether it's high, low, or somewhere in between, the fact of the matter is just the same: this country faces a huge, complicated, and scary energy problem.

Not only is it huge, complicated, and scary, it's daunting for most of us because it's nearly impossible to understand unless you're some kind of full-time energy wonk. We all recognize symptoms such as soaring gas prices and pricey home heating oil. But when it comes to the disease—what's causing the country's energy problems and how to cure them—that's another story entirely. Confused R Us is more like it.

It's not that the information isn't out there. Type "energy policy" into Google, and you'll get tens of millions of results. Some of the stuff is biased and manipulative, and some of it is downright harebrained. But there is plenty of information from smart, responsible, fair-minded experts who really know this issue and want to help. Unfortunately, when most of them talk, they might as well be speaking Greek (or Urdu or Basque). All of a sudden, it's "peak oil" and "strategic reserves" and what's happening at the New York Mercantile Exchange. Want to know about possible solutions? Prepare yourself for treatises on "carbon sequestration," "hydraulic fracturing," and the promise of "photovoltaic cells." Not only is it hard to understand, it can often be incredibly boring. Say the words "energy policy" often enough, and you could undermine the entire profit structure of the sleeping pill industry (not to mention upsetting the little counting sheep from the mattress company).

Even so, we're convinced that the vast majority of Americans do want to know what can be done to ensure that the country has safe, reliable energy at nonstratospheric prices. Nearly all of us want to protect the planet, the economy, and our way of life. But sorting out the country's choices is tough—almost as tough as facing up to them.

PULLING THE STRANDS OF THE PROBLEM TOGETHER

The first step is to pull the far-flung pieces of this debate together in one place. There's the energy issue with its assorted disputes over OPEC, oil company profits, speculation in the energy markets, and how to reduce the country's dependence on imported oil. Then there's the environmental debate on how to reduce the damage human beings do to the planet—global warming, carbon dioxide emissions, carbon footprints, pollution, that sort of thing. And finally, there's the economic fallout when the competition for energy heats up and supplies start getting tight. That's what happened in the first half of 2008: skyrocketing gas prices, major airlines in financial free fall, and businesses of all sorts hurting because families have to spend more on gas and electricity and have less money left to spend on everything else.

These three issues—energy, environment, and the economy—are all intertwined, and some experts say the country could at some point get itself trapped in a "perfect storm" with all three problems coming to a head at once. In the movie *The Perfect Storm*, George Clooney sailed his fishing boat right into "the storm of the century" in the North Atlantic. It was spine-tingling to watch, but he killed himself and his crew doing it.

For the United States, sailing right into the triple threat we face on energy is an equally bad idea. Unfortunately, we've been backing ourselves into an energy corner for a good forty years or more, and the country simply has to start making some responsible decisions beginning right now.

SIX REASONS TO ACT

Let's get started with the basics. Here are six reasons the United States needs to get its energy act together. There isn't much debate about these facts, but there is plenty over what to do about them.

The country is using more energy than ever. The U.S. population is growing, we're doing more driving than we did a few decades ago, and we have more energy-eating devices than ever. Just think of all the technology we've added to our lives in the last generation: computers, microwaves, DVDs, MP3 players, a cell phone for every member of the family. Since 1949, the total amount of energy used in the United States has tripled, and we're not done yet.[1] Americans are expected to consume about 25 percent more electricity over the next twenty years.[2]

Other people around the world are using more energy too. According to the best estimates, global demand for energy is expected to grow by 45 percent over the next two decades,[3] and specialists who study energy supply and demand are asking some fairly scary questions about where all that energy is going to come from.[4] The Earth's population is growing, but it's not just that. Roughly a quarter of the world's people don't even have electricity yet. To improve their lives, developing countries will need a lot more energy than they have today—73 percent more by 2030, based on expert projections.[5] What's more, people in developing countries such as China and India are now becoming prosperous enough to want to live the way we do. They like cars and TVs and computers and warm houses filled with gadgets. Don't get us wrong; this is a good thing. People everywhere naturally want the comforts money can buy. But it also means more competition for energy, higher prices, possible shortages, and potential environmental damage on a scale we've never seen before.

We're relying on forms of energy that will eventually run out. In the last 150 years or so, human beings used about 1 trillion barrels of oil. Some experts say we could use up another trillion in about thirty years.[6] There is a big debate (which we cover later) over whether and how quickly humankind is running through the Earth's supply of oil, but some of the predictions are getting uncomfortably close.[7] We can certainly look for more, and no doubt we'll find it, but some experts worry that we're rapidly using up the oil that's relatively easy to access. There's also some concern that in North America at least, the remaining natural gas supplies are in locations where it will be difficult (and costlier) to extract them.[8]

As supplies get tighter, energy costs more. Energy prices tend to go up and down, and perhaps the only bright spot in a recession is that energy prices generally fall because people use less of it (factories and businesses closing, fewer people driving and traveling, etc.). But whatever the price of oil may be when you read this book, the overall trends are just not in our favor. As recently as 2004, many analysts thought oil would stay at about $30 a barrel for the next decade.[9] In 2008, the average for the year was just under $100 a barrel, even though prices fell dramatically in November and December in the economic downturn.[10] Many experts believe oil prices will start rising again when economies worldwide begin to recover and the competition for oil heats up again.[11] But it's not just oil. Prices for natural gas more than doubled between 2002 and 2008.[12] Prices for the uranium used for nuclear power also doubled between 2006 and 2008.[13] When more people want more of something, and it's not lying around all over the place, and it takes a long time to find it and put it into usable forms, prices tend to go up. Since we consume energy when we make, ship, and use everything from big-screen TVs to Pop-Tarts, the price of fuel spills over into our entire economy.

The U.S. energy supply system is shaky. In 2007, the United States used about 7.5 billion barrels of oil and imported 58 percent of it.[14] Unfortunately, a fairly large portion of the world's oil reserves lie in some of its most unstable regions (like the Middle East) and in the hands of potentially unstable governments (like Nigeria and Venezuela). That means a whole host of things can go wrong—embargoes, war, revolution, terrorist attacks on pipelines. As drivers learned after Hurricanes Katrina and Rita, a natural disaster that disrupts shipping and refining can upset the balance too. But it doesn't require an international crisis or an act of God to give us problems with our energy supply. Our electric power grid and oil and natural pipelines are aging. The older they get, the more they're prone to failure.

No one fools around with Mother Nature. Hurricanes strong enough to disrupt shipping (such as Hurricane Camille in 1969) can also disrupt the oil supply. Photo courtesy of the National Oceanic and Atmospheric Administration

Oil, coal, and gasoline—the kinds of energy we use most—can be harmful to the Earth and everything living on it. Burning fossil fuels such as these causes pollution and acid rain, and according to most scientists, it contributes to global warming—at least the way we do it now. But it's crucial to make an important distinction. The United States has made good strides reducing air pollution and acid rain because of the Clean Air Act and better technology in our cars and power plants. Unfortunately, we haven't done nearly as much to reduce the emissions that contribute to global warming, so that's the next big challenge. The problem of global warming is "unequivocal," according to the Nobel Prize–winning Intergovernmental Panel on Climate Change.[15] It's "clear," according to the American Association for the Advancement of Science.[16] There's "a growing scientific consensus," according to the U.S. Congressional Budget Office.[17] ExxonMobil, which at one time supported research by climate change skeptics, now runs ads highlighting the need to address the problem.[18] Even people with lingering doubts seem to accept the main point. The grumpy but always intriguing Charles Krauthammer considers himself a global warming "agnostic," yet he says he "believes instinctively that it can't be very good to pump lots of CO_2 [carbon dioxide] into the atmosphere."[19] Okay, that's it. We have way more than a quorum.*

WAITING FOR THE PERFECT SOLUTION

So there we have it: six solid reasons to rethink the country's energy policy. Yet despite the dangers, the United States has been stuck in first gear for years now. We haven't upped domestic oil production much because of environmental concerns. We haven't moved vigorously to develop alternatives because we frittered away our time arguing about whether global warming is real. We invest in the scientific research and technology to help solve our energy problems only when energy prices get high. Since we're so fond of big cars and big houses filled with labor-saving devices, we've barely scratched the surface on conserving the energy we do have. We've spent several decades just sitting around waiting for some perfect, cost-free, "please don't ruffle our feathers" solution to come down from the sky.

That hasn't happened yet, and it won't. But there are lots of reasonable options most of us can live with. There are lots of innovations that could help us down the road if we get to work on them now. This is going to be a long

---

* There are lots of better people than us to explain the science of climate change. The U.S. Environmental Protection Agency maintains a Climate Change page chock full of clear explanations from reliable sources (www.epa.gov/climatechange/index.html). NASA's page, Earth's Fidgeting Climate, is another good place to go (http://science.nasa.gov/headlines/y2000/ast20oct_1.htm).

That said, while we have a quorum, complete unanimity is nearly unheard of in any human endeavor. The prominent physicist, Freeman Dyson, has written extensively about his doubts. Also, if you're curious about what the disbelievers say, take a look at *The Deniers* (Richard Vigilante Books, 2008) by Lawrence Solomon.

battle, though, and what happens in Washington, D.C., is just the beginning. What we need is a state-by-state, power-plant-by-power-plant, business-by-business, car-by-car, house-by-house rethink.

The United States desperately needs to get a move on in making these choices, so that's what this book is about: laying out the challenge and explaining the options. Let's have the debate, make some decisions, and get on with it.

We started the chapter with a little bit from the Rolling Stones: "You can't always get what you want." If you've got Mick on the brain now, maybe you remember that the song goes on to say that you might be able to "get what you need." Not such a bad description of our situation right now. If the United States can get its act together on a reasonable, long-term energy policy, we may well be able to get what we need. If not, let's hope the next song that comes to mind isn't "Paint It Black."

Playing the Blame Game

If you haven't followed the energy issue much, you might wonder why the country has made so little progress and what we've been doing instead. Well, we've had quite a lot of fun playing the energy blame game. It's a common human proclivity. We don't want to blame ourselves or accept that unpleasant facts and impersonal forces limit our options. If something bad is happening, it must be because some villain somewhere is making it happen for his or her own evil reasons. Maybe it's some completely bald guy with a vaguely foreign accent, a secret lair, and an employee motivation program that relies heavily on a shark tank. You know, like Blofeld in the James Bond movies.

So for your consideration, we've put together some nominees for the title of all-time energy evildoer. You may see some of your favorites here. If you happen to see yourself on the list, don't worry. We all have plenty to feel guilty about.

Okay, it's your turn to choose. Who's the worst energy villain of all time? The nominees are:

OPEC. The Organization of the Petroleum Exporting Countries has been giving the United States fits in the energy department almost since it was founded in Baghdad on September 14, 1960. Saudi Arabia, Venezuela, Kuwait, Iran, Indonesia, and the other eight OPEC members[20] produce about 45 percent of the world's crude oil.[21] We generally want them to produce more so that prices won't be so high.

The oil companies. In one of the worst economic downturns in history and with oil prices plummeting toward the end of

2008, ExxonMobil still had record U.S. profits of over $45 billion for the year. Even oil companies that couldn't keep up with Exxon still tended to do better than most other businesses in a supremely tough business year.[22] Six in ten Americans say the U.S. oil companies deserve "a great deal" of the blame for the country's energy problems.[23]

Speculators. Given what wild-eyed investors did to the U.S. economy during the Internet bubble of the 1990s and the home mortgage bubble not so long ago, it's not surprising that many Americans blame speculators in oil markets every time gas prices go up. Congress fumes about them; regulators look into it. As columnist Paul Krugman wrote in 2008 when gas prices topped the $4 mark: "The welcome mat is out for analysts who claim that out-of-control speculators are responsible."[24]

"Tree-huggers." According to the Energy Department, there are about 18 billion barrels of oil in the country's outer continental shelf that are currently off-limits to drilling.[25] Environmentalists have long battled to bar offshore drilling, along with oil and natural gas exploration and drilling in federally-owned lands. For many Americans, it's the environmentalists who are keeping the country's energy policy in virtual shackles. Critics say environmentalists unreasonably fight proposals to tap domestic oil offshore and in Alaska and are hindering the country's ability to use much of its coal and natural gas. Plus much of the environmental movement opposes nuclear power, even though it doesn't cause global warming.

People who drive big SUVs and Hummers (or used to). Not only do they drive these mega-gas-guzzlers, they had the nerve to make them fashionable, at least for a good number of years. Back then, even if you weren't so keen on having a big, hefty vehicle yourself, you felt like you had to buy one to avoid being squashed like a bug if you crashed into one on the highway. SUV sales more than tripled between 1988 and 2007.[26] Some people blame the automakers, who in turn say they were making the kinds of vehicles buyers wanted. Take your pick.

NIMBYs, or people who fight having oil refineries or nuclear plants or electric lines or windmills anywhere near them. The last significant new oil refinery in the United States opened in 1977.[27] Proposals to build new nuclear power plants frequently spark local opposition. Residents organized to fight windmills off Cape Cod, and many experts worry that local opposition will get in the way of rebuilding the nation's electric grid. Communities often organize to make sure these sorts of things aren't located in their neck of the woods. It might be okay somewhere else, but not here. It's the dreaded phenomenon known as NIMBY. That's "not in my backyard" (and not to be confused with Gumby).

Politicians who keep gas prices low (and the voters who expect them to do it). Economists often have a favorite energy villain: politicians who have kept U.S. gas prices among the lowest in the developed world. Over the last thirty years, apart from a few spates of gas price unpleasantness, Americans have pretty much been able to drive as much as we want in whatever kind of car we like and go as fast and as far as our little hearts desire. For many experts, the obvious answer, maybe the only answer, to reducing Americans' dependence on oil is to raise the price of gasoline.

And then there's the rest of us.

Yes, we all know we should change our ways, but many of us also feel like we're caught between a rock and a hard place. We love and need our cars. This whole country, with its interstate highway system and suburban lifestyle, was designed with cars in mind. Outside of a few places such as New York and Chicago, it would be a real schlep for most of us to get to work without our cars. (Did you ever try to take the bus anywhere in Albuquerque, New Mexico? It takes forever.) So the moment talk turns to energy, most of us begin worrying that the big "they" will start piling on whopping gas taxes (it already costs enough, thank you very much) and forcing us into itty-bitty cars for the daily commute. That's when we start looking for someone else to blame. Surely this must be someone else's fault.

That, in a nutshell, is the blame game. Take a look back over our list. Is your top villain there? Want to add someone—maybe some elected officials you're not too fond of? Maybe you think we were unfair to include good old so-and-so?

Actually, we hope that's the case, because that's precisely our point. When it comes to energy, nearly all of us have already decided who the good guys and bad guys are. It's a very convenient way of looking at things—seething over OPEC is a lot easier than thinking seriously about what the United States should do because OPEC probably isn't going to go away anytime soon. Shaking your fist at the oil companies is easier than taking the time to understand the country's energy problems and thinking through what our choices are. Even better, you don't have to admit your own role in the equation. It lets you off the hook.

But the time for the blame game is over. (And it has been such fun, hasn't it?) Unfortunately, it's way too simplistic to think that the kind of energy challenge the United States faces can be solved by taking out a couple of bad guys, even if you're convinced they have it coming.

The Great Energy Debate Pop Quiz

The energy issue is very confusing, and frankly, most of us will never catch up with the experts on all the details. Still, there are some basic facts that are good to know. Do you know them?

True or false? When it comes to global warming and air pollution, nuclear power is one of the most dangerous forms of energy.

Not true. The accidents at Chernobyl and Three Mile Island left lots of people worried about nuclear plant safety, but if you're worried about climate change, nuclear power is one of the *least* dangerous forms of energy we have. Generating electricity from nuclear power releases virtually no carbon dioxide (the major greenhouse gas) into the atmosphere, and it doesn't cause air pollution either.[28] Small amounts are emitted during mining and processing the uranium (you need uranium for nuclear power) and in other related activities, but it's nearly impossible to do anything from start to finish without releasing some greenhouse gases. Experts say the carbon dioxide released in these associated activities puts nuclear power roughly on a par with wind or hydroelectric power.[29] Like every form of energy we've discovered so far, nuclear power does have drawbacks, but global warming isn't one of them. The big drawback to nuclear power is that the leftover waste, the spent fuel, has to be stored very, very carefully, and it lasts a really, really long time. Even so, nuclear power is widely used in Europe and Japan, and despite the controversies about it, it supplies 19 percent of electricity in the United States—enough electricity to keep air conditioners, TiVos, and iPods going in California, New York, and

Texas.[30] Scientists are working on other ways to dispose of nuclear waste, including recycling it into the nuclear power plant itself, but the problem hasn't been solved yet.[31] See Chapter 9 for a more complete discussion of the pros and cons of relying more on nuclear power, including the safety issues.

True or false? ExxonMobil, BP, and Chevron control nearly half of the world's known oil reserves.

Not even close. In fact, none of the big multinational oil companies we complain about so often even makes the top ten list. So who's controlling the lion's share of the world's oil reserves?

The national oil company of Saudi Arabia (Saudi Aramco) has the most oil reserves, followed by the national companies

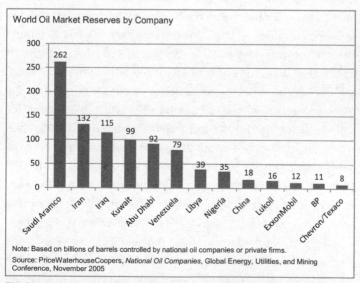

World Oil Market Reserves by Company

Note: Based on billions of barrels controlled by national oil companies or private firms.
Source: PriceWaterhouseCoopers, *National Oil Companies*, Global Energy, Utilities, and Mining Conference, November 2005

The famous names in the oil business don't actually control most of the world's oil anymore. National oil companies now have most of the oil.

of Iran, Iraq, Kuwait, Abu Dhabi, Venezuela, Libya, Nigeria, China, and Lukoil, the largest oil company in Russia. Exxon-Mobil comes in at number 13, BP at 15, and Chevron/Texaco at number 20.[32] Congress likes to have the corporate heads of the major oil companies appear in hearings so our elected representatives can have their fifteen minutes of fame asking tough questions about gas prices, but in many ways, the big multinationals such as ExxonMobil have much less control over the country's oil situation than they once did.

Which country is guiltiest when it comes to releasing greenhouse gases into the atmosphere, the United States or China?

It's a trick question because, frankly, the United States and China are running neck and neck for worst greenhouse gas polluter in the world.[33] There are several ways to look at this, and none of them exactly puts the United States in the clear. Global warming is caused by the *accumulated* greenhouse gases in the atmosphere (carbon dioxide and its brethren). Since the United States got a head start (we started using large quantities of coal about the time of the Civil War), our country alone is responsible for about 29 percent of the total accumulated gases, compared to just 8 percent for China.[34] Then there's the per person measure. In 2005, each American gushed out about 20 metric tons of carbon dioxide, compared to about 5 metric tons for each person in China.[35] But China has a billion more people than we do, and they are building and manufacturing and transporting like crazy there now. If the average Chinese person begins emitting greenhouses gases at the same rate as the average American, it will just wallop the environment. As of now, China is producing about 21 percent of the world's carbon dioxide emissions.[36] Bottom line? The United States and China need to stop pointing fingers at each other. Both our countries really need to get with the plan.

Who sets the price for a barrel of crude oil?
A. OPEC
B. Oil companies such as ExxonMobil and Saudi Aramco
C. The U.S. Department of Energy
D. The New York Stock Exchange
E. None of the above

The answer is none of the above. The price is actually set by bidding, buying, and selling on major commodities trading exchanges in New York, London, and Singapore.[37] These are different from the stock market, but they operate in a similar way. Basically, traders buy and sell all day at the best price they can get, which is why the price for a barrel of oil goes up and down so much and generally changes daily.[38] That doesn't mean that OPEC and other oil producers have no impact on prices. As OPEC itself puts it, member countries "do voluntary restrain their crude oil production in order to stabilize the oil market and avoid harmful and unnecessary price fluctuations."[39] In other words, they calculate how much they're willing to pump based on the price they want to get. One of the disputes about oil that erupts from time to time is the degree to which the OPEC countries are producing as much as they can or whether they are holding back. (You can read more about what affects the price of oil in Chapter 6.) On the other hand, since the oil is theirs to extract and sell, it's also fair to ask whether, from their point of view, they should produce as much as they can as quickly as they can, or whether they want to preserve some of their countries' natural resources for the future.

What percent of the world's known oil and natural gas reserves are in the United States? A. About 20 percent B. 10 to 20 percent C. 5 to 10 percent D. Less than 5 percent

In area, the United States is the world's third-largest country; only Russia and Canada have more territory than we do.[40] Unfortunately that doesn't mean we control a substantial share of the world's oil and natural gas reserves. According to 2008 estimates, the United States has about 2.4 percent of known world oil reserves and about 3.6 percent of natural gas reserves.[41] These are figures for the "known" or "proved" reserves—that is, geologists actually know the stuff is there—so more exploration could definitely up those numbers a tad. However, as we mentioned earlier, many experts believe that the remaining U.S. supplies of both oil and natural gas are in less convenient places and less convenient forms (such as tar pits). That means they'll be costlier to extract. Just to make your day, would you like to know that Iran, which is tiny in comparison with the United States, has more than four times as much oil[42] and natural gas as we have?[43]

True or false? As long as global warming doesn't increase world temperatures more than 5 or 10 degrees, the effects will be easily manageable.

Not according to the climatologists who worry about global warming. In 2007, the UN's Intergovernmental Panel on Climate Change summed up the judgment of scientists worldwide predicting that average global temperatures will rise 3.5 to 8 degrees by the year 2100.[44] It sounds minor. After all, most of us would be hard pressed to say whether the temperature was 70, 75, or 80 on a nice spring day. But sustained changes like this over time cause glaciers to melt and sea levels to rise. People living near water, especially poor ones in poor countries, can be displaced, and miserable, disease-carrying microbes can flourish. It changes what crops you can grow where, which can cause serious economic and social upheaval. In 2008, the U.S. government released a report summing up the scientific consensus on what climate change could mean here in the United States.[45] Among the conclusions: it is "very likely" that "abnormally hot days and nights and heat waves" will be more frequent, increasing the number of people who die from heat-related causes, especially the elderly, frail, and poor. The report warned that "climate change can also make it possible for animal-, water-, and food-borne diseases to spread or emerge in areas where they had been limited or had not existed." Lyme disease and West Nile virus are two examples mentioned.[46] As we said before, there are a lot of good reasons to revamp the country's energy policies, and global warming is only one of them. But if you'd like to see exactly what the scientists are worried about, you might want to check out NASA's "Eyes on the Earth" interactive global time line showing the changes in sea levels and the polar ice cap that scientists are already observing. It's at www.nasa.gov/multimedia/mmgallery/index.html.

Suppose People in China Really Start Living the Way We Do

I can feel it when they come into the shop. The whole family chooses the car together.... After they take it home, they get up several times every night to see if their cars are okay.

—*Xing Chuang, Huizhongtong Automobile Trading Company, China*[47]

For the last half century, the American standard of living has been the envy of the world. Throughout the 1950s, '60s, and '70s, Americans lived longer and more comfortably than just about any human beings since the beginning of recorded time, even the richest ones. There's a scene in the movie *Lion in Winter* where King Henry II of England (Peter O'Toole) has to break the ice in his washbowl before washing his face in the morning. Think about it: He's the king, he lives in a castle with an army of servants, and he's *still cold in the morning.* Are you? Your nice little three-bedroom, two-car-garage American house has plenty of hot water. Good solid appliances. Plenty of electronic gadgets around the house. A big car in the driveway—certainly one, probably two.

It's the American dream—at least the material part of it. And the Europeans and the Japanese have their version too. Compared to the developing world, we're all at the top of the heap. But now we're being joined by an expanding roster of countries. China, India, South Korea, and Indonesia are growing with amazing speed. In nine respects out of ten, that's the best thing that could happen. A world with a handful of very rich countries and lots of poor ones is not likely to be a safe, peaceful place. As other countries become more prosperous, there's no reason to expect that their citizens won't aspire to the same kinds of comforts and possessions we have. They're not monks. Why wouldn't they?

And that brings us to the Chinese and the issue we're tackling in these pages—how people use energy, where they get it, and what it does to the planet. Experts think it will take at least a couple of decades before most people in China enjoy the kind of material wealth middle-class Americans do. But the Chinese are definitely on their way. Their overall economy has been growing at *six times* the rate of the United States' economy.[48] Many Chinese families are starting to have incomes that allow them to buy the kinds of things we have and enjoy—things that use a lot of energy.

Since there are about four times as many people in China as there are in the United States, it's fairly obvious that worldwide energy use is going to soar. China will likely overtake the United States as the world's largest energy consumer in the next few years, and their overall energy use will double by 2030.[49] That means more competition for the energy that's out there, and, as long as we're mainly relying on fossil fuels, more environmental damage as well. The impact is potentially huge—enormous, massive, gigantic, gargantuan, really, really big. (It's a handy feature, that thesaurus button. Wonder whether the Monty Python crew relied on it when they created their dead parrot routine.)

Here are some guesstimates about what could happen if the Chinese people begin to live the American lifestyle:

Population	Total Energy Use	Oil Consumption	Electricity Consumption	Cars	Greenhouse Gases (CO_2 Emissions)*
United States: Roughly 300 million	In 2006, the United States used 99.9 quadrillion Btus; China used 73.8 quadrillion Btus.	In 2007, the United States used about 20.7 million barrels per day, while China used 7.6 million barrels daily.	The United States used 3,817 billion kilowatt hours of electricity in 2007; China used about 2,529 billion kilowatt hours, but remember, China has a billion more people.	In 2002, the United States had about 812 motor vehicles per 1,000 people, for a total of 234 million; China had about 16 per 1,000 people, for a total of 20.5 million.	In 2007, Americans, doing all the things we need and like to do, released about 5.9 billion metric tons of CO_2. Right now, the Chinese don't drive as much; they don't use as much electricity; they don't have all our appliances—at least not yet—so even though they have many more people, the Chinese released about 6 billion metric tons of CO_2 that same year.
China has about 1.3 billion people. Yep, they have a billion people more than we have.	If the Chinese begin using energy the way we do, they would be using 432.9 quadrillion Btus per year.	If the Chinese begin using oil the way we do, that would be about 89.7 million barrels a day.	If the Chinese started using electricity the way we do, that would be nearly 17,000 billion kilowatt hours a year.	If the Chinese start owning cars at the rate we do, that would be over 1 billion cars.	If the Chinese started spewing out greenhouse gases at the same rate as Americans, that would be about 25.6 billion metric tons of CO_2 a year. That's a frightening amount—almost as much as the entire world put out in 2003.

SOURCES: Energy Information Administration, World Resources Institute, and Dargay, Gately and Sommer, *Vehicle Ownership and Income Growth, Worldwide: 1960–2030*, Department of Economics, New York University (2007), and the United Nations Human Development Report, 2007–2008.

* We'll mercifully skip the details on how experts calculate the amount of greenhouse gases various countries release. Frankly, it's an estimate; it has to be. The one we're using is made by the U.S. Energy Information Administration.

So what does it mean for energy and the environment if the United States and China just keep on going the way we're going? Using our handy little thesaurus button and the Monty Python motif, it's terrifying, fear-provoking, worrisome, petrifying, and alarming. Unless we all want to wind up as dead parrots, we've got to find a way of dealing with this.

CHAPTER 2

GROUNDHOG DAY, OR HAVEN'T WE SEEN THIS MOVIE BEFORE?

Phil: Do you have déjà vu, Mrs. Lancaster?
Mrs. Lancaster: I don't think so, but I could check with the kitchen.

—*Groundhog Day*

The United States has been around the block a couple of times on energy. In fact, you might think the country was caught in one of those repeating time loops that trapped the starship *Enterprise* from time to time. Or maybe we're like Bill Murray living the same sequence of events over and over again in *Groundhog Day*. The country keeps running into energy problems, and we do the same things again and again. Oil and gas prices shoot up. Voters get upset. Politicians say we absolutely, positively need to change the way we get and use energy. We make a few adjustments here and there—both in the country's overall policies and in our own personal habits. Then a couple of years later, we go back to our same old ways. At least Bill Murray changed his line of attack after hearing Sonny and Cher sing "I Got You Babe" on his radio alarm for the umpteenth time.

Since your authors believe that if you don't at least glance at history, you might be doomed to repeat it, we'll take a quick look at some of America's previous energy adventures in this chapter. If you can't stand history, don't toss the book across the room yet. Our trip is so speedy you may not even notice. If you do like history, you might enjoy looking over our Energy Chronicles on pages 32–36. You might also want to check out our online appendix at www.whoturnedoutthelights.org, where you can find some good books and websites on the subject.

There's no test coming up here. We just want to underscore the point that when it comes to energy, the United States has been there and done that a couple of times. Then we let ourselves backslide. This time has to be different, and we want you to understand why.

DON'T BE FUELISH!

Here's a snapshot of the country's most famous (or infamous) energy crisis. In the early 1970s, Arab oil producers and their OPEC partners embargoed shipments to the United States in retaliation for American support of Israel in the Yom Kippur War. All of a sudden, the country was short on oil and gas, and at the height of the crisis, gasoline was rationed. (Yes, younger folk, such a thing actually happened in the United States.) In some areas of the country, you could only buy gas on odd- or even-numbcred days depending on whether your license plate ended in an odd or even number. There were even rumors that unscrupulous two-car families lucky enough to have odd-ending plates on one car and even-ending plates on the other would switch the plates around so they could buy gas any day they wanted. Human iniquity knows no bounds when it comes to getting gas for the family car.

The handwriting isn't so great, but the message was clear enough. Once upon a time, the United States ran short on gas. Courtesy of the U.S. Department of Energy

Over the next several years, the government took other steps to try to get a grip on the situation. Congress lowered the speed limit to 55 mph to save gas. We got an "energy czar" and added a Department of Energy to the government. We had year-round daylight savings time for a while and a "Don't Be Fuelish" public service campaign. In 1975, Congress passed legislation requiring automakers to improve fuel efficiency for passenger cars, and this last step in particular triggered some bona fide change.[1] Over the next decade, American carmakers were able to double miles per gallon ratings for autos, from 13.5 up to 27.5. In 1979, when the shah of Iran was overthrown and radicals seized the U.S. embassy in Tehran, there was a second oil shock that sent prices shooting up again and kept the country focused on the issue.

But the nation's commitment to the plan petered out as prices went down once more and memories of gas lines receded. By the mid-eighties, automakers had successfully persuaded Congress to ease up on fuel efficiency standards, especially for light trucks, a move that prompted then-Chrysler chairman Lee Iacocca to comment: "We are about to put up a tombstone, 'Here lies America's energy policy.'"[2] Iacocca wasn't wrong. Between the mid-eighties and 2007, the typical American-made vehicle bulked up by about 900 pounds and doubled its horsepower. Naturally, auto fuel efficiency dropped.[3]

By the 1990s, the 55 mph speed limit was on life support,* and high-flying trucks and SUVs whizzed past drivers pokey enough to obey the 70 mph speed limit. Year-round daylight savings time is a thing of the past, although not everyone agrees that this idea is actually helpful in reducing energy use.[4] The Department of Energy is still with us (and they have a slew of good statistics on their website that your authors appreciate very much), but the energy czar is just a regular old cabinet secretary, and not the most prestigious one at that. Can you name the secretary of energy? Can you name any secretary of energy from the last decade?†

Unfortunately, America's boom-or-bust mind-set on energy just keeps on keeping on. When gas prices are high, we start looking seriously at ways to conserve. When gas prices drop, we revert back to our gas-guzzling, "my car is bigger than your car" way of thinking. In 2002, when gas was less than $1.50 per gallon, GM sold 71,000 Hummers and its executives were contemplating expanding the Hummer line.[5] By 2008, when gas suddenly hit the $4 mark, the country's love affair with cars getting 16 miles a gallon began to fade.[6]

---

* Congress repealed the national 55 mph speed limit in 1995.

† For the record, and for what it's worth, the current secretary is Dr. Steven Chu. Previous secretaries were Samuel W. Bodman (2005–8), Spencer Abraham (2001–5), Bill Richardson (1998–2001), and Federico Peña (1997–98) (www.energy.gov/news/4757.htm).

In 2009, GM announced that it was selling off the Hummer brand.[7] Fuel-efficient cars were suddenly fashionable. More people began carpooling and using public transportation. Even bike sales went up.

OIL, COAL, GAS–IT'S BEEN A GREAT RIDE

There's no arguing that oil, coal, and gasoline have given this country a great ride. They fueled the Industrial Revolution and made transcontinental and international travel practical. At the end of World War I, there were 26 million mules and horses in the United States, used mainly to get people and things around. Now we have about 250 million vehicles, pretty much putting the mules and horses out of the transportation business.

But this country's reliance on these fossil fuels— especially the way we use them now—is beginning to be a treacherous choice. In the past, we've been able to hunker down during an oil crisis and then go back to normal once the worst is over. Now there's ample evidence that we need to change our ways for good.

HERE'S WHY THINGS ARE DIFFERENT NOW

BACK THEN

The country's "energy crises" were caused by relatively brief disruptions somewhere in the system.

- In the 1970s, war in the Middle East led OPEC to cut oil shipments. That led to gas rationing and other ills here, but after a while, things returned to normal.
- In the early 1980s, the price for a barrel of oil jumped again,[8] because the United States placed an embargo on Iranian oil after Iranian militants took American diplomats hostage there.[9] The hostages eventually came home, and world oil markets returned to normal.
- In the early 1990s, prices jumped up when Iraq invaded Kuwait, and the United States and its allies launched the Persian Gulf War to force Iraq back within its borders. The United States won that war handily, and sooner than you can say "jackrabbit," SUVs were all the rage back home.
- In 2005, Hurricanes Katrina and Rita pushed gas and oil prices up for a spell because some refineries were knocked out by the weather. But with time, this problem was solved too.

The point is that all of these situations, however troubling and unpleasant, were temporary.

TODAY

The conditions driving the country's energy problems are almost certain to last a very long time. Let's recap:

- We're competing with many more people for the energy that's available. The world's population is approaching seven billion. By 2050, it's expected to be well over nine billion.[10] Every single person on the planet needs and uses energy. Unless we develop some good alternatives, they're likely to be using oil, coal, and gas, the same fuels we're so dependent on. We don't have to tell you what that means for prices (but we will anyway: They'll be going up). And we don't have to tell you what it means for global warming. It's not going to be good.

- There are truly astonishing levels of economic growth in China, India, and elsewhere. These countries need massive amounts of energy for their factories and transportation. As these countries and others become more prosperous (which is generally considered progress), the people living there will start buying cars and refrigerators and microwaves and computers. All these things use energy.

- As we mentioned earlier, there's a growing consensus among experts that humans are beginning to use up most of the oil that's easy to get to. (We go over the details on this in Chapter 6.) Remember that alarming little factoid from Chapter 1: It took more than a century for human beings to use up the first trillion barrels of oil, but experts say we'll burn our way through the second trillion in about thirty years or so. We don't want to be alarmist, because the world is not actually going to run out of oil in our lifetimes. But chances are that it's going to get tougher and tougher to find it.

DON'T STOP THINKING ABOUT TOMORROW

What does all this mean for the good old USA? Basically, it means that unless we get our act together, the country could wind up in a permanent energy crisis that goes on for decades.

The trouble is that every time oil and gas prices come down a bit (and they go up and down all the time), the United States boomerangs back to the good old days: all the energy you want at prices you can easily afford. It's quite possible that prices for oil and gas will stay relatively low in the next few years due to the global economic shake-up. Not to be too gloomy about it, but there's nothing quite like a worldwide recessionary downturn to bring energy prices down. In nation after nation, people use less energy because there's less manufacturing, less travel, less of just about everything. When that happens, oil and gas prices generally drop. But when the economy rebounds and people start doing all the things they do when times are good, prices tend to go right back up again.[11] It's just that weird old supply and demand thing.

But maybe oil and gas prices will stay low for other reasons. Maybe someone will discover a nice new batch of oil someplace—not enough to solve the world's long-term problem, but enough to bring prices down for a while. This doesn't let us off the hook.

We need to get a sound plan together, and this time, stick with it even when the pressure seems to be off. Besides, there are so many options for solving this problem. Maybe we won't all agree on the best ones, and no doubt we'll make mistakes along the way. But the most damaging mistake of all would be to assume that our energy problems are temporary and will fade away with time. That would be a truly gigantic error.

America's Energy Chronicles, or How We Got to Where We Are Today

1859: Former railroad conductor Edwin Drake hits oil in Titusville, Pennsylvania. Drake's discovery leads to the Pennsylvania oil rush.

1882: New York City generates electricity from a coal-fired plant based on a design by Thomas Edison.

1908: Henry Ford brings his Model T to market. The Model T could run on gasoline or ethanol.

1914: Standard Oil of California opens a string of thirty-four gas stations up and down the West Coast.

1936: Congress passes the Rural Electrification Act to provide electricity for people living in agricultural areas.

Thomas A. Edison. Photo cour
Prints and Photographs Divisio
Library of Congress

1946: The Atomic Energy Commission is established to develop peaceful uses for nuclear energy.

1957: The first commercial nuclear power plant opens at Shippingport, Pennsylvania.

1963: Congress passes the first Clean Air Act.

1969: More than 3 million gallons of crude oil spill into the Pacific Ocean near Santa Barbara, California, from a Union Oil drilling platform, leading to new environmental legislation.

1970: The first Earth Day

1973: After the Yom Kippur War breaks out between Israel and a coalition of Arab countries, the United States faces an oil embargo, and gas is in short supply.

Photo courtesy of the Franklin D. Roosevelt Library, Public Domain Photographs, 1882–1962

1977: President Jimmy Carter signs the Clean Air Act Amendments designed to reduce auto emissions and improve fuel efficiency.

1978: Congress passes a "gas guzzler" tax of up to $7,700 per vehicle.

1978 Oil prices soar when the shah of Iran is overthrown.

1979: An accident at Pennsylvania's Three Mile Island nuclear plant is the most serious in U.S. history. The actual release of radiation is small, and no deaths are directly attributed to the accident, but it generates broad public fears about the safety of nuclear power.

1986: A much more serious accident at the Chernobyl nuclear plant in Ukraine forces the immediate evacuation of more than 100,000 people. It is estimated that more than 4,000 cancer-related deaths may

Photo courtesy of the U.S. Department of Energy, National Renewable Energy Laboratory

eventually be attribut-
able to the accident.

1988: A NASA scientist, Dr.
James Hansen, testi-
fies to Congress on
the dangers of green-
house gases.

1989: An oil tanker, the
Exxon Valdez, runs
aground in Alaska in
Prince William Sound,
spilling more than 11
million barrels of oil in the largest oil spill in U.S. history.

Photo courtesy of the U.S.
Department of Energy

1991: Oil prices jump after Iraq invades Kuwait and the
United States and its allies launch the Persian Gulf
War.

1992: The Environmental Protection Agency begins its Energy Star program to promote energy efficiency.

1997: The Kyoto Protocol to limit greenhouse gases emissions is created, and over the next several years, more

Courtesy of the National Oceanic and Atmospheric Administration

than 150 nations agree to sign it, but the United States is not one of them.

2001: Enron, once the world's largest electricity and natural gas trading company, files for bankruptcy after a series of scandals.

2002: General Motors introduces the Hummer.

2005: Motorists in some parts of the country face long gas lines after U.S. refineries shut down temporarily in the wake of Hurricane Katrina.

2007: Former Vice President Al Gore and the United Nations' Intergovernmental Panel on Climate Change win the Nobel Peace Prize.

2008: The price for a gallon of gasoline tops $4 for the first time.

2009: The Obama administration proposes major investments in energy research and "green jobs," and increases mileage requirements for cars. General Motors goes through bankruptcy and sells its Hummer division to China's Sichuan Tengzhong Heavy Industrial Machinery Company.

Photo by Warren Gretz. Courtesy of the U.S. Department of Energy, National Renewable Energy Laboratory

CHAPTER 3

GIVING THE VOTERS WHAT THEY WANT

I was watching that movie *Mad Max*, you know, that movie where gas is so precious that people are killing each other for a few gallons. It was set in the future—I believe it was August.

—*Jay Leno*

Some things are Washington's fault: a national capital filled with some 15,000 lobbyists, billions spent on earmarks, and a $12 trillion national debt and counting. Maybe we should have been paying more attention, but it's hard to argue that these things are really what most Americans want from government.

In the case of energy, though, the American people aren't quite so blameless. In fact, you could say that over the last few decades, elected officials have given Americans exactly what they want on energy and the environment—a no-muss, no-fuss continuation of the status quo.

Sure, importing all that oil and watching polar bears clinging to melting ice floes are upsetting, but Americans have been very slow to demand action and/or accept genuine changes in our energy habits. Besides, why should we rouse ourselves to tackle the energy problem when there are so many other people we can blame—OPEC, the oil companies, the automakers, the Chinese, the Republicans, the Democrats? Why should we have to change?

Despite our slow start, there's some evidence that the country is starting to inch its way to a more vigorous energy policy. But as the debate becomes more concrete—as elected officials in Washington and the states start grappling with more and more real proposals—it's fair to ask whether their decisions will reflect the common good (not the "special interests") and whether they'll really be thinking about the long-term health of the country (not the next election cycle).

UNDER THE INFLUENCE?

There's no question that there are plenty of people telling Congress and the administration what the country should do about energy. The advice and the lobbying come from different directions—traditional energy companies, alternative energy companies, environmental groups, states with economies that depend on different kinds of energy, trade unions, and more. Since nearly all major American industries keep an eye on what is happening in government, it would be pretty surprising if oil, natural gas, and coal companies were any different. In 2008 alone, the oil and natural gas industries spent about $129 million on lobbying and utility companies spent even more, about $157 million.[1] What's more, these industries regularly make campaign contributions.

However, we have to be clear-eyed about exactly how much influence these traditional energy industries really have in the nation's capital. When it comes to campaign contributions for members of Congress, the oil and gas business came in at only number seventeen among the top political donors in 2008, with electric utilities close behind at number twenty[2] The automotive industry, which is now relying on taxpayer-financed loans for its survival, is even further behind, at number thirty-four.[3] The biggest givers are those who list their profession as "retired," and lawyers and law firms.

Moreover, the political influence of the energy industry has been at least partially countered by the political influence of environmentalists. It's not a dollar-for-dollar match—far from it—but the environmental movement has been persuasive and persistent in mounting its case. And decision makers in Washington and elsewhere are often disposed to listen.

In 2008, environmental groups spent more than $17 million on lobbying efforts, with groups such as the Environmental Defense Action Fund, Nature Conservancy, Defenders of Wildlife, National Wildlife Federation, Trout Unlimited, and the U.S. Climate Action Partnership among the leaders.[4] The movement also contributes to candidates spending more than $4 million in 2008.[5] That is, of course, chicken feed compared to what oil and gas companies gave (about $35 million).

In a democracy, it's a given that individuals, groups, and, yes, even profit-making companies will do what they can to make sure elected officials understand their point of view (within the bounds of the law, of course). That's part and parcel of the way our democracy works. Despite the money that oil, gas, and utility companies have spent in Washington over the last several decades, they haven't always gotten their way.

According to Lindsay Renick Mayer of the Center for Responsive Politics, an organization that tracks political contributions and lobbying, "Campaign contributions don't always get the oil industry desired results."[6] She points out that "four of five of Big Oil's most favored candidates—all Republicans—lost their re-election races in 2006, despite hefty campaign contributions from oil and gas employees and PACs."[7]

"WE HAVE MET THE ENEMY AND HE IS US"

Cartoonist Walt Kelly came up with this phrase a half century ago in his comic strip *Pogo*. (If you haven't seen it, *Pogo* was the *Doonesbury* of the fifties.) The line certainly applies here, because it's worth asking whether lobbying and campaign contributions from corporate and environmental interests really explain why the country has done so little to address its energy problems over the last several decades. Maybe. Partly. Still, it's hard to wiggle out of one stubborn truth: A major cause of the nation's energy and environmental gridlock has been the mind-set of the American public.

Despite all the evidence that the country needs to change its energy and environmental policies, public opinion has not rallied to demand action, and Americans have not coalesced around a set of solutions. In fact, it's safe to say that for many years, as long as gas prices were moderate, the country's energy and environmental problems barely crossed voters' minds.

It's not that most Americans don't recognize the dangers of our current energy policies, if someone asks about them. Based on the Energy Learning Curve, a 2009 public opinion survey conducted by our organization, Public Agenda, about seven in ten Americans say they expect that oil prices will rise over the long run because of growing demand and diminishing supplies. In fact, eight in ten Americans say they're at least somewhat worried that the economy is too dependent on oil and that our oil habit might pull us into another Middle East war.[8]

The angst and anxiety evaporate when the topic of a gas tax comes up. The vast majority of Americans just don't want anything to do with a gas tax. Solid majorities reject even a relatively modest tax (40 cents a gallon) targeted specifically to helping the country become energy-independent.[9]

The global warming side of the picture isn't exactly a call to action either. Most Americans don't think global warming will pose much of a threat to their way of life in their lifetime, based on surveys by Gallup.[10] They don't dismiss the threat entirely (only 13 percent say we can keep doing what we're doing now). Most just don't see global warming as an urgent problem (only 34 percent think it requires "additional, immediate, drastic action").[11] Read it and weep, Al Gore.

A FOG OF CONFUSION

Energy and environment are complicated issues (we're going to use about 300 pages trying to sort them out), so maybe it's not surprising that Americans haven't reached a consensus on the need for action or the preferred solutions. But the Energy Learning Curve survey suggests that the lack of knowledge about some important and undisputed facts is unnerving. After all, it's hard to make good decisions when you've got basic facts all wrong.

- Half of Americans think that since we've reduced smog in the United States, we've already gone a long way to reducing global warming.[12] Sadly, no. Less smog is good, but you can't see carbon dioxide. Reducing the visible pollutants that cause smog is not helping out much on the global warming front.

- Four in ten Americans think that if we drill offshore and in Alaska, we wouldn't need to import foreign oil.[13] It would probably help, but the United States used about 7.5 billion barrels of oil in 2007. Experts think that at most there's about 16 billion barrels of "technically recoverable" oil in Alaska—a little over three years' worth.[14] The majority of Americans also think the United States has at least 10 percent of the world's oil. It's less than 3 percent.[15]

- Surprisingly large numbers of Americans think that nuclear power and solar power cause global warming. There are pros and cons to every form of energy, but nuclear and solar are almost entirely in the clear on the global warming front.

   Part of the problem is that the worst of the world energy crisis and the related environmental dangers are still down the road a piece—or at least they seem that way to people preoccupied with their day-to-day lives. Plus the issues are complicated, and the truth is that we're all going to have to change some of our expectations and habits to make things work. But you wouldn't know that from listening to much of the political discourse over the years. Rather than explaining the issues, many of our leaders have seemed more interested in fashioning slogans to persuade people that they have the answer—answers that won't cost anyone anything much. No wonder Americans are confused.

   To make good decisions, we need to get some basic facts under our collective belt, and that's basically what the rest of the book is about—giving you the background you'll need. It's worth taking time to learn something about energy because we're going to be working on this problem for a good long time.

   In Ken Burns's multipart documentary, *The Civil War,*

the narrator begins each of the nine episodes with the same phrase: "The Civil War was fought in ten thousand places." The point is crucial to understanding what the country went through when the North fought the South. It wasn't just the big battles with the famous names—Gettysburg, Cold Harbor, Shiloh. The war was everywhere. The North's victory—and the South's defeat—resulted from events that took place all across the country.

The country's escape from its energy dilemma will be much the same. It won't happen because we do one thing; it will happen because we do many things in many places. And it will take considerably longer than the Civil War, which lasted four years; the four decades of the Cold War might be a better guess. But if this country survived challenges as daunting as the Civil War, two world wars, the Great Depression, and more, surely we can get our act together to solve our energy problems. It will take resolve, but no shooting is required.

CHAPTER 4

SEEMED LIKE A GOOD IDEA AT THE TIME, OR HOW THREE FLAWED IDEAS COULD GET US OFF TRACK

Many a beautiful theory was killed by an ugly fact.
                                    —*Thomas Henry Huxley*

For a long time, energy was a second-tier issue. Politicians talked about gas prices, of course—whenever they started creeping skyward, they immediately became a hot political topic. More farsighted debate on the country's overall energy situation? Well, that was mainly for wonks and experts. Recently, however, that's started to change.

In the 2008 elections, even candidates for dogcatcher seemed to be talking up their plans for quick action to reduce our reliance on foreign oil and keep energy prices low. And why not? Quick, cheap, and made in the USA—these ideas are very popular with most people (and maybe even their pets). They sound good at first blush, so one of the surprising things about the energy issue is how wrongheaded those ideas could turn out to be.

Knowledgeable experts and decision makers actually raise very unsettling questions about all three issues. Most don't think the country can or should be completely independent of imported energy. Most see worrisome problems with energy getting too cheap. And virtually all emphasize that it will take a long time—a very long time—to make significant changes in the way the United States gets and uses energy. Change won't happen overnight.

Let's take a look at each of these ideas separately and examine why something that sounds so good might not be such a great idea after all.

FLAWED IDEA #1: THE UNITED STATES SHOULD BECOME ENERGY-INDEPENDENT

Politicians have latched on to the public's anxieties about foreign oil by inserting lots of references to "energy independence" into speeches and interviews. For a country that got its start with the Declaration of Independence, it's easy to see why this phrase has an attractive ring to it, but scholars at major policy organizations across the political spectrum, ranging from the Council on Foreign Relations to the Brookings Institution and the Cato Institute, all say actual independence is a pipe dream. What's more practical and smarter, most say, is "energy security," meaning working to get the United States less dependent on foreign energy sources—not closing the tap entirely.[1] Why is energy independence such a flawed idea?

- **It would be nearly impossible to do it anytime soon.** Almost all our cars, trucks, and planes run on petroleum-based fuels, and we import nearly 60 percent of the fuel we use, so this isn't something that can be fixed in a year or two. The U.S. Energy Information Administration is predicting that the country can reduce its imports over time, but they still believe that unless we change our current policies radically, we'll be importing about 40 percent of our oil in 2040.[2]

- **Not every foreign energy source is a problem.** Most of the enthusiasm for the idea of energy independence is based on the understandable concern that the United States is getting oil from the Middle East, as well as from other potentially unstable countries such as Venezuela and Nigeria. Some of the countries we get oil from certainly are unstable, hostile, or both. But our biggest oil supplier is Canada. On *South Park*, they famously sing "blame Canada" for all our problems. But out here in the nonanimated world, Canada is not likely to cut us off, and we certainly don't gain anything from being independent from them. Our second-biggest supplier is Mexico, which has its problems, but it is also too close—and too closely intertwined—with the United States to make isolation feasible.

- **It might not actually protect us from swings in oil prices.** There's a global market in oil, and it's going to continue whatever we do. U.S. energy companies buy and sell their products and services worldwide, and a big run-up in oil prices would just encourage them to export more rather than keep it at home. The kinds of laws, taxes, and tariffs needed to stop that might ignite a trade war. Experts point out that countries such as Britain, which exports oil thanks to its North Sea offshore rigs, gets hit just as hard by world oil prices as we do. "No country, not even a net exporter, can stop the world and get off," says Pietro S. Nivola, an energy scholar at the Brookings Institution.[3]

- **In fact, it might actually cost more.** One main reason we buy oil on global markets is because on any given day petroleum may well be cheaper somewhere else. Relying purely on domestic production means less competition, and less competition usually means higher prices.[4]

The far better way, most experts say, to make sure the United States has energy security is to have lots of different sources of energy supplied by lots of different companies in lots of different places. It's that "don't put all your eggs in one basket" thing, because even a U.S. basket full of U.S. eggs might have problems—such as hurricanes in the Gulf of Mexico knocking out a chunk of U.S. oil production and refining. Then we might be happy to import a little oil from Canada up north.

FLAWED IDEA #2: CHEAPER ENERGY IS BETTER

It's easy to see how $100 fill-ups for the family car and soaring home heating bills play havoc with the economy. When people spend more on energy, they don't have as much left to spend elsewhere. When businesses spend more on energy, they have less for employee raises and for creating and bringing new products to market. When transportation is expensive, it adds to the costs of virtually every kind of product you can name. All of this can really drag the economy down. No surprise, then, that an oil price spike was a major factor in tipping over the house of cards that was the global financial system in 2008.

But most of us don't think as much about what happens when energy gets really cheap. On the surface, it might seem as though plummeting prices for oil, natural gas, and other kinds of energy should be fabulous for everyone. In fact, really cheap energy has some worrisome downsides.

- Really cheap energy means there's no incentive to change, either by developing energy-efficient products or by looking for alternatives to fossil fuels. If something's cheap and plentiful, there's no incentive to conserve it or to find an alternative. Cheap gas leads normally reasonable people to buy cars that practically inhale the stuff. If energy is too cheap, inventors and manufacturers don't have as much incentive to develop new products and technologies that could help us conserve energy and eke as much as we can out of what we do use. The same goes for new energy sources. When investors and entrepreneurs believe that prices for traditional energy sources are going up and staying high, they see an opening for wind and solar, and for ideas like electric cars. When fossil fuel prices are very low, they know that their products will be at a serious cost disadvantage. That means their offerings will be less likely to succeed, and the investors stand more chance of losing money on them.

- **Really cheap oil prices mean possible shortages and even higher prices in the future.** There are a couple of issues here. One is that when oil prices are low, oil companies don't spend as much time looking for more of it. Investors and energy companies aren't as eager to explore and put money into drilling, processing, and transporting it unless they are going to get a good return on their money. This is especially true for projects that are pushing the envelope by looking for oil in out-of-the-way places and in "unconventional" forms.* (The same is true for natural gas, by the way.) Moreover, when oil prices get very low, oil-producing countries (and groups of countries, such as OPEC), start cutting back production in an effort to keep prices from falling more. Just as sure

---

* See for example, "Crisis Hits North American Oil Investment," Reuters, October 24, 2008, about exploration projects that were postponed or scaled back when oil prices dropped in the last half of 2008, or Clifford Krauss, "As Oil and Gas Prices Plunge, Drilling Frenzy Ends," *New York Times*, March 15, 2009, www.nytimes.com/2009/03/15/business/15drilling.html.

as shooting, say the energy experts, that will lead to rising prices in the future. And since there's a long lead time involved, it's not easy to go out and start drilling and exploring again. If the demand for oil gets way ahead of what's being pumped out of the ground, then prices can really soar. With worldwide demand growing, there may not be enough for everyone at prices they can actually afford.

- **Really cheap oil prices could cause instability and violence in some parts of the world.** A number of oil-producing countries depend on oil for nearly all of their income, and when prices drop sharply, those governments could be in deep trouble. Venezuela and Iran, in particular, need fairly high oil prices in order to keep their country's economies afloat, and Russia could also face some serious financial shortfalls. A lot of Americans find these countries irritating or even threatening, so it's hard to feel sorry for them. But an economic collapse in any of them might lead to unrest, danger and hardship for the people who live there, and even more troublesome leaders or factions taking over.[5]

- **Really cheap energy lulls us into thinking everything is fine.** This could be the worst drawback of all. In fact, this is what has happened in the United States repeatedly. When oil and gasoline prices get sky high, we start getting our act together—conserving, moving to more energy-efficient ways of doing things, looking at and investing in alternatives to fossil fuels. When prices fall, we slide right back into our energy-wasting ways.

There are some economists who argue that all this isn't a problem. When and if prices go up and stay up, we *will* conserve, we *will* find alternatives. But as we've said, the problem is lead time, and that brings us to another trouble-some assumption people often make.

FLAWED IDEA #3: WE CAN CHANGE THE ENERGY PICTURE QUICKLY IF WE JUST DECIDE TO DO IT

For a country that loves fast food and FedEx, the national psyche may need an attitude adjustment when it comes to energy. Most of us don't think much about energy until it's an emergency, and then we want it fixed now. But the real world doesn't always jump in line just because we'd like it to. Most ideas for transforming the way the United States gets and uses energy are going to take a while, and "a while" often means decades. Why do things take so long? Here are some reasons:

- **Many ideas still need a lot of work.** Almost every day you hear about some fantastic new idea for solving the country's energy problems—new technologies, new sources of energy, new products that use energy much more efficiently. Some of them might very well work out nicely eventually, but just because it's on TV doesn't mean it's ready for prime time. Most of these things aren't sitting on a shelf somewhere waiting for someone to take them down. Take hydrogen fuel cells, for example. The technology dates back to the Apollo space program in the 1960s, and when and if it's ever ready, fuel cell technology would be a great source of electricity with little pollution. But all the experts say it's going to take years of research before you can make these light enough and practical enough for cars. You've heard people moan about how "we can put a man on the moon, but we can't do X," adding their own pet peeve in place of the X. Well, fuel cells prove the point: they can take a man to the moon, but they're not ready to take your minivan on a cross-country road trip.

- **Just having the technology is not enough.** It's got to make sense economically. Money is a big factor, no doubt about it. Yes, there are people out there who'll buy more expensive products because they're "greener," but realistically most customers won't buy an electric car if it costs significantly more than a regular one. The economic climate makes a difference too: people are more willing to try something different when gas is at $4 than they are when it's at $1.50.

- **Changing over takes time.** Let's say car companies have worked out the kinks, and fuel cell cars are competitively priced. The problem is that most typical paycheck-to-paycheck people don't buy a new car at the drop of a hat, even if they'd like to (and they definitely don't do it when the economy is struggling). The typical car on the road is more than nine years old because lots of drivers hold on to their car until it's basically done for (or they trade it in and someone else drives it until it's basically done for).[6] Technology changeovers take time. Not everyone jumps in right away. Think of all the people still using VCRs and dial-up Internet service. There people may seem strange to you, but they do exist. Believe it.

What is the estimated time of arrival for some of the big ideas for getting the United States out of its energy mess? Here are some guesstimates to ponder. We'll use 2010 as our start date.

Five years: Solar power should be "unsubsidized and competitive" by 2015, according to goals set by the Department of Energy, but the U.S. Government Accountability Office

(GAO) thinks the goal is too optimistic.[7] Right now solar power is four times as expensive as other sources of electricity, and provides less than 1 percent of the country's electricity.

Ten years: That's the time it would take for oil leases in the Arctic National Wildlife Refuge to begin to boost domestic oil production, according to estimates from the Energy Information Administration (part of the Department of Energy).[8] The estimate covers the time needed for exploration, drilling, and building the infrastructure needed to support it, but doesn't include time that might be needed to settle lawsuits filed by local residents or environmental groups opposed to the plan.

Twenty years: One-fifth of the country's energy could come from wind power in about two decades, according to estimates from the Department of Energy. That's if the country triples the number of wind turbine installations by 2017. Logically enough, DOE calls its report *20 Percent Wind Energy by 2030*.[9]

Twenty-three years: This is the time it took the Tennessee Valley Authority to complete work on Unit 1 of its Watts Bar nuclear plant near the Chicamauga Reservoir in eastern Tennessee. Construction started in 1973 and the unit began commercial operation in 1996, the last nuclear plant to go online in the United States. Of course, it didn't actually take that long to build it; what got in the way was all the regulatory back-and-forth, construction delays, and protests.[10] The physical process of building a nuclear plant takes between six and ten years.[11] Assuring everyone that it's a safe idea and a good one can take considerably longer. Work on a second reactor at Watts Bar was suspended in

1988 and resumed in 2007.[12] If it goes online in 2013 as
scheduled, its contruction and development will come in at
an even forty years.[13]

Forty years: Burning coal to generate electricity releases
large quantities of carbon dioxide into the air, which contributes to global warming. But by 2050, according to the
Electric Power Research Institute (EPRI), 40 percent of coal-
burning power plants in the United States could have new
technology that removes and stores the carbon dioxide so it
doesn't escape and add to our climate change problems.[14] It
would require aggressive research and development, and
many environmentalists think the money could be better
spent on alternatives such as wind and solar power, but
prominent scientists and engineers say it's technologically
possible and both the United States and the world have an
awful lot of coal. It wouldn't just happen overnight.*

What's the bottom line? Some ideas that sound good
are not only unrealistic but actually get in the way of getting things done. Energy independence, eternally cheap energy, and quick fixes are implausible at best, dangerous
distractions at worst. On the other hand, if we look at what
we really can do—and how quickly we can do it—the
United States can make real progress.

---

* More on this possibility in Chapter 7.

A Message to Our Readers

In the rest of the book, we'll get down to the nitty-gritty. But before moving on to the details, we want to introduce a theme—a quandary, if you will—that winds through just about everything we have to say from now on.

Basically, the United States is in trouble because we rely so heavily on fuels that will be in increasingly short supply and endanger the Earth to boot. Unfortunately, these same fuels (oil, coal, natural gas) are the lifeblood of our economy and way of life. Going cold turkey isn't realistic, so here's the dilemma: should the United States concentrate mainly on working with the traditional forms of energy we already have, or should we start to move away from them and into alternatives as quickly as we can? Let's take a quick look at both points of view.

WE SHOULD WORK WITH WHAT WE HAVE

The emphasis here is on finding more energy within the United States so we don't have to import as much. Many experts argue that the United States has reasonably large supplies of oil and natural gas but that we haven't done a very effective job of getting them out to the market. They argue that the environmental dangers of drilling, mining, and pipelines have been exaggerated, especially given the technological improvements the industry has made in recent years. Many are interested in natural gas. It's a fossil fuel, so using it does release some greenhouse gas, but it's a lot cleaner than most others. Proponents back research to reduce the harmful side effects of fossil fuels, especially coal. Plus, if U.S. scientists and companies were the first ones to do it really well, we could probably sell that technology worldwide and make ourselves some good money on it.

What's more, many of those who want to put the emphasis on the traditional fuels we have here in the United States fear that moving too quickly to costlier alternatives would harm the American economy overall. Plus, it would likely hurt consumers in regions where electricity is mainly generated by fossil fuels and devastate communities in states where the oil, coal, and gas industries are major employers.

LET'S MOVE AWAY FROM FOSSIL FUELS AS QUICKLY AS POSSIBLE

For other Americans, the number one goal is to cut back on oil, coal, and natural gas as rapidly as humanly possible because of what those fuels do to the planet. Experts on this side of the debate believe that the environmental dangers of fossil fuels have, if anything, been *under*estimated. To them, drilling for more oil and natural gas in the United States and building more refineries and pipelines are risky, irresponsible distractions. Rather than feeding our addiction to oil, coal, and natural gas, they say, we should move full throttle toward renewables, including wind, solar, and geothermal. Some propose something like a Manhattan Project (the collective effort of the scientists who developed the atom bomb) or the Apollo Project (to land a man on the moon) that would bring the nation's top scientific and engineering talent to work on alternatives and renewables and ways to up efficiency. They also recommend redoubling efforts to conserve, especially while we're still using fossil fuels. For this group, the economic danger lies in not moving quickly enough. They believe the real threat to our prosperity is in standing around twiddling our thumbs while scientists and inventors in other countries develop the alternative energy products and services that will be urgently needed worldwide.

MIXERS AND MATCHERS

These two paths are not absolute opposites, of course. You could expand oil, coal, and natural gas production in the United States *and* launch a Manhattan Project on alternatives *and* put money into demonstration projects to eliminate carbon dioxide emissions from coal. T. Boone Pickens, the oilman turned wind entrepreneur, is a mix-and-matcher. He supports expanded use of natural gas to replace gasoline, but he's also proposing that the United States set a goal of getting 20 percent of its energy from wind power. The former CEO of Royal Dutch Shell, Jeroen van der Veer, is so concerned about the mounting worldwide demand for energy that he urges full throttle ahead in all directions: "It is not a matter of choice, do we do coal, or oil, or nuclear?" he said in 2008. "The world will need everything, including biofuels. You name it."[15]

But there are tensions. Where should we put most of our money? What should we emphasize? What should we discourage? What's the best balance given that changing the country's energy habits will take time? What should we do about industries and their employees and communities that take a hit in the changeover?

And what are we willing to do ourselves? Are we willing to pay more at the pump, drive a smaller car, pay more for electricity, have more nuclear plants in our state, or new electric lines put up in our neighborhood? When the time comes, what trade-offs are we willing to make?

We'll return to these differing points of view many times in this book. Our aim is to help you think about the arguments on both sides so you can come to your own conclusions.

CHAPTER 5

THE BASICS: TEN FACTS YOU NEED TO KNOW

If it weren't for electricity, we'd all be watching television by candlelight.

—*George Gobel*

Most Americans don't have the slightest idea where their energy comes from, or how much they use. They do know how much they pay, of course, and whom they hand the money over to, either at the gas station or on their utility bill. And most know the United States imports most of our oil. Beyond that, quite a few of us would flunk Energy 101.

To think sensibly about solutions, however, we need a good grip on some basic facts, such as where our energy comes from, how we use it, and what the implications of that are. So settle back for the big picture.*

---

* The main source for this chapter is the U.S. Energy Information Administration's Annual Energy Review, the government's statistical report covering all kinds of energy use going back to 1949. All the gory details are at www.eia.doe.gov/aer/.

1. THE UNITED STATES IS CONSUMING MORE ENERGY THAN EVER, AND THERE'S NOTHING NEW ABOUT IMPORTING IT

There's absolutely no doubt that the United States uses a great deal more energy than it used to, and that much of it comes from abroad. As a nation, the United States uses three times as much energy as it did fifty years ago, and overall energy consumption has jumped more than one-fifth since 1990. After all, the population has grown, and so has the U.S. economy. What's more, energy production here in the United States has slowed since 1990, so the gap between how much we produce and how much we use has been getting wider. Naturally, that means imports are growing as well.

This chart, which lays out the country's rising appetite for energy and growing reliance on imports, is what makes a lot of people blurt out, "Oh my God, we've got to get energy independent now!"

But have another look and consider this: When was the last time the United States could cover all its energy needs? Think Sputnik. Think Elvis. Think enormous Chevys with

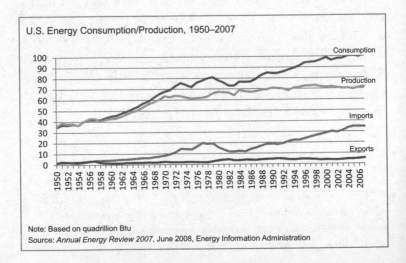

U.S. Energy Consumption/Production, 1950–2007

Note: Based on quadrillion Btu

Source: *Annual Energy Review 2007*, June 2008, Energy Information Administration

huge tail fins. In reality, we haven't been self-sufficient in energy since 1957. Importing energy isn't something that crept up on us recently. Two whole generations have grown up in the half century since then. If this was easy to solve, it wouldn't have gone on this long.

2. AT LEAST HALF THE ENERGY IS USED BY BUSINESS AND INDUSTRY

In the United States, we often have a tendency to act as though gasoline were the only kind of fuel we need to worry about and as though cars are the be-all and end-all of the United States' energy problem. We also have a tendency, encouraged by both energy companies and environmentalists, to believe that the country's energy challenge is mainly about what we do as individuals. We're constantly bombarded by people telling us that if we just turned off the lights (or at least switched lightbulbs), we'd be okay. People's personal habits are certainly important—more about that

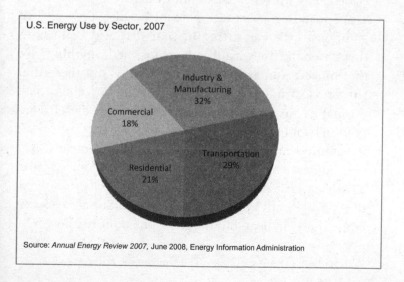

U.S. Energy Use by Sector, 2007

Industry & Manufacturing 32%

Commercial 18%

Transportation 29%

Residential 21%

Source: *Annual Energy Review 2007*, June 2008, Energy Information Administration

later—but they're far from the whole story. At least half the country's energy use is for industry, manufacturing, and commerce.

How we use energy breaks down into four big categories, and each of those categories has a distinct need and sometimes a distinct kind of energy.

- **Industrial** accounts for nearly a third of all energy use, the largest single sector. In addition to factories, this covers agriculture, mining, and construction. Mostly this involves powering machinery as well as the heating and cooling needed to actually make things (heating the forges at a steel mill, for example, or refrigeration at an ice cream company).
- **Commercial** covers stores, offices, government buildings, and, for reasons known only to the government bureaucrats who calculate these things, both sewage plants and "institutional living quarters" such as college dorms.* That's 18 percent of the total, primarily drawing from electricity with some use of natural gas.
- **Transportation** covers 28 percent of all energy use. In addition to the car sitting in your driveway right now, this covers airlines, railroads, ships, and anything else that moves. And aside from a few trains out there that run on electricity, almost all of this (96 percent, at last estimate) runs on some form of oil: gasoline, diesel, aviation fuel, and the like.
- **Residential** means just what you'd think it means: private homes. That accounts for 21 percent of energy use, mostly natural gas (for heating) and electricity (for everything else, mostly appliances and air-conditioning). Some homes have oil heat, but they're in the minority now.

---

* We don't know about you, but our dorms weren't *that* messy.

What this means is that many of the most important choices about how the country as a whole uses energy—and what kinds we use—are made by business, utilities, and government. It's not going to be enough for us all to behave better in our personal lives. We need a much more comprehensive plan.

3. THE UNITED STATES OVERWHELMINGLY GETS ITS ENERGY FROM FOSSIL FUELS

The country uses energy produced from oil, nuclear power, natural gas, and many other sources, but the big three, accounting for over 80 percent of the country's energy consumption, are all fossil fuels.[1] Petroleum supplies nearly 40 percent of the country's overall energy needs; natural gas contributes another quarter, and coal a little over a fifth. As

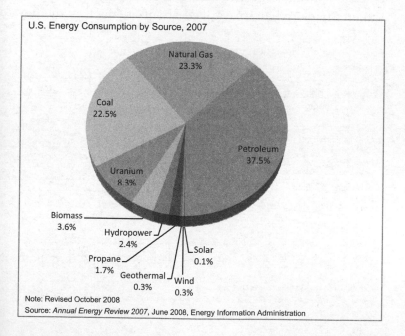

U.S. Energy Consumption by Source, 2007

Natural Gas 23.3%

Coal 22.5%

Petroleum 37.5%

Uranium 8.3%

Biomass 3.6%

Hydropower 2.4%

Propane 1.7%

Geothermal 0.3%

Wind 0.3%

Solar 0.1%

Note: Revised October 2008

Source: *Annual Energy Review 2007*, June 2008, Energy Information Administration

we'll keep saying until the cows come home, all forms of energy have advantages and disadvantages (and the middle section of this book is devoted to a guided tour of the pros and cons of each option), but it's important to keep the big picture in mind. The United States relies heavily on fossil fuels to cover its energy needs, and fossil fuels have two major drawbacks—using them pumps huge amounts of greenhouse gases into the atmosphere, and sooner or later they'll run out.

4. GREENHOUSE GAS EMISSIONS IN THE UNITED STATES COME MAINLY FROM TRANSPORTATION AND GENERATING ELECTRICITY

Since we use cars, trucks, and airplanes to get around, and since they nearly all run on some form of petroleum, most Americans aren't too surprised to learn that transportation is a big source of greenhouse gas emissions here in the United States. What isn't talked about as much is the de-

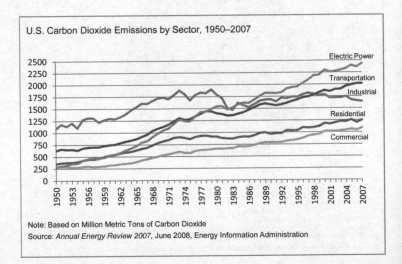

U.S. Carbon Dioxide Emissions by Sector, 1950–2007

Note: Based on Million Metric Tons of Carbon Dioxide
Source: *Annual Energy Review 2007*, June 2008, Energy Information Administration

gree to which the way we produce electricity contributes to global warming.

Think about it. You know what you're getting when you fill up your tank with gas, but what does your power company use to generate electricity?

The short answer is, almost anything. Nationwide, the most likely bet is coal, which not only accounts for half the electrical power in the country but also is hardly ever used for anything else, at least in the United States.[2] Natural gas and nuclear power have significant shares too, each accounting for about one-fifth of the national power supply.[3]

Despite all the talk about wind and solar power, they constitute a very small portion of the country's current electricity supply. Both are growing industries, and they have many advantages, but right now very few homes rely on electric power supplied by them. The fact is, 70 percent

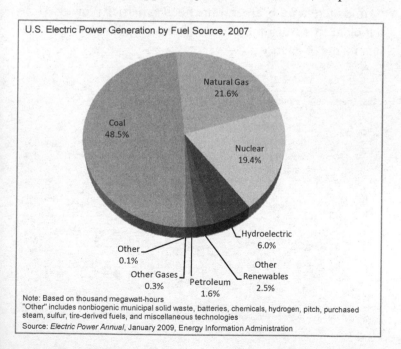

U.S. Electric Power Generation by Fuel Source, 2007

Natural Gas 21.6%

Coal 48.5%

Nuclear 19.4%

Hydroelectric 6.0%

Other 0.1%

Other Gases 0.3%

Petroleum 1.6%

Other Renewables 2.5%

Note: Based on thousand megawatt-hours
"Other" includes nonbiogenic municipal solid waste, batteries, chemicals, hydrogen, pitch, purchased steam, sulfur, tire-derived fuels, and miscellaneous technologies
Source: *Electric Power Annual*, January 2009, Energy Information Administration

of our electricity comes from coal or natural gas, and both release carbon dioxide (although natural gas produces significantly less than coal).

What kind of power you're getting depends a lot on where you live. Electrical power is still very much a regional business, and decisions by local utilities and state regulators play a big role. The states along the Pacific Coast, for example, get more than 40 percent of their power from hydroelectric dams and most of the rest from natural gas and nuclear power; coal hardly factors in at all. Midwestern states, on the other hand, depend heavily on coal, some for more than 70 percent of their electricity.[4]

5. WHEN TIMES ARE BAD OR PRICES ARE HIGH, WE USE LESS ENERGY

Sounds like a no-brainer, but this is worth exploring. Have a look at this consumption chart showing how we use various forms of energy.

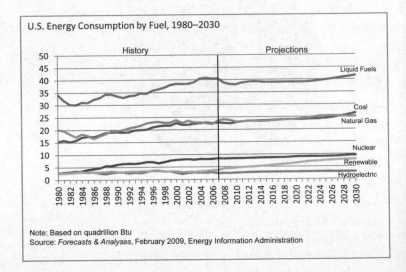

U.S. Energy Consumption by Fuel, 1980–2030

Note: Based on quadrillion Btu
Source: *Forecasts & Analyses*, February 2009, Energy Information Administration

The overall trend is up, up, up. But notice there are dips, especially in how much oil we use: a big one in the late seventies and early eighties, and smaller ones in the early nineties. So what happened?

Those dips line up nicely with recessions—which makes perfect sense. When the economy slows down, businesses cut back on production, fewer people have jobs to commute to, and people don't fly as much, so energy use goes down. This is what happened in the recession that began in 2008. When the economy gets rolling again, energy use goes up. And since the economy has grown more often than it's shrunk over the past forty years, the consumption line usually goes up.

The other factor here is price. When energy gets more expensive, people use less of it. Again, this makes perfect sense. During the oil price spikes of the seventies and early eighties, the country got serious about conservation, changing building codes, upgrading utility plants, and setting fuel economy standards for cars for the first time. This is another pattern we saw in 2008, when oil reached more than $100 a barrel.[5] People drove less, took public transit more, and otherwise cut back.

Prices, of course, don't stay put. In the 1980s fuel got cheaper, given some major shifts in the world market. Using dollars adjusted for inflation, gas prices in the early 1980s were hovering around $3 per gallon.[6] By 1986, the cost had dropped by half, and prices didn't rise appreciably for any substantial length of time until 2005. When gas is comparatively cheap and plentiful, we don't bother conserving it. So, as a nation, for a long stretch of time, we basically stopped worrying about energy and the price of gas.

We'll be talking about the question of price a great deal in this book since we tend to conserve and look for alternatives only when fossil fuels start getting expensive.

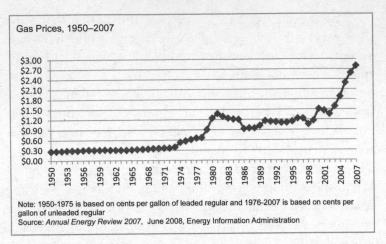

Gas Prices, 1950–2007

Note: 1950-1975 is based on cents per gallon of leaded regular and 1976-2007 is based on cents per gallon of unleaded regular
Source: *Annual Energy Review 2007*, June 2008, Energy Information Administration

Consequently, an important part of the policy debate is about whether the government should deliberately make fossil fuels more expensive, perhaps by raising the gasoline tax or imposing a "carbon tax" or permit fee on anything that uses fossil fuel.

6. THE UNITED STATES IS MORE ENERGY-EFFICIENT THAN IT USED TO BE

It's important to understand the various reasons U.S. energy consumption has been rising over the last fifty years, and one reason is that the country itself has grown. There are more than 300 million people in the United States today, and that's 125 million more than in 1960.[7] A larger population means more energy use, even if the amount each person uses stays the same. We're a prosperous country compared to many others, and in general, people use more energy as their income rises (they can afford to buy cars and TVs, maybe even several of each—you get the picture).

However, the amount of energy each of us uses indi-

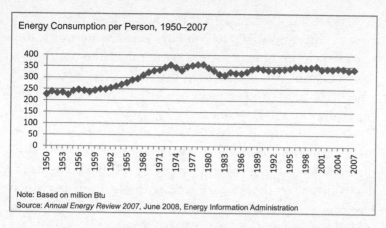

Energy Consumption per Person, 1950–2007

Note: Based on million Btu
Source: *Annual Energy Review 2007*, June 2008, Energy Information Administration

vidually has actually leveled off since the late 1970s. It's a piece of good news to relish: The country is more energy-efficient than it was thirty years ago. Cars get (slightly) more miles per gallon, although nearly everyone agrees we could do better.[8] You can buy all those household appliances with the Energy Star logo on them. Many industries and commercial enterprises have put a lot of effort into being more efficient in how they use energy, because that keeps their costs down, which helps them keep their profits up.

In 1980, energy consumption per person ran at about 345 million Btu of energy a year—that's counting everything, from filling up your car with gas to running factories.* By 2005 it was still 340 million Btu a year. If you plot this on a chart, it actually looks really, really boring—none of the sharp peaks and valleys that chart makers love. So despite our love of cars and gadgets, we are actually using about the same amount of energy per person as Americans were using two decades ago. The snag is that there are far more of us.

---

* *Btu* stands for "British thermal unit," a standard measure for the energy content of fuels.

There's another way to think about how the country uses energy, and that's to compare our energy use to the size of the economy, or, as the economists like to say, as a percentage of gross domestic product. GDP is the total amount of goods and services produced in a given year, the sum total of the U.S. economy.

Obviously the U.S. GDP is bigger than that of most other countries—even those with more people (such as India, which has three times as many people but is not as industrialized) or more space (for example, Canada, which has a huge expanse of land filled with caribou). The question of how much the United States spends on energy as a percentage of American GDP is essentially an economist's complicated way of answering a simple question: how much bang are we getting from our energy buck? Amazingly, energy spending as a percentage of GDP has been going down.[9]

Still another way to look at energy efficiency is to calculate how much fuel we're using to produce a dollar's worth of goods and services. And it's been going down. In 1970, it took nearly 18,000 Btu of energy to produce a dollar's worth

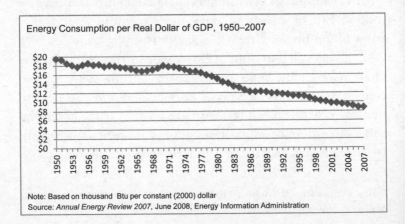

Energy Consumption per Real Dollar of GDP, 1950–2007

Note: Based on thousand Btu per constant (2000) dollar
Source: *Annual Energy Review 2007*, June 2008, Energy Information Administration

of stuff.* By 2006, we only needed around 8,700 Btu to create a dollar's worth of goods and services. Partly, of course, that's a reflection of the twenty years or so of cheap energy we had after 1985. During most of that time, the U.S. economy was growing and worker productivity was rising as well, so the country was getting more economic benefits from our energy spending.†

7. WE DRIVE MORE, AND WE OWN MORE GADGETS

Even though transportation accounts for less than a third of the country's energy use, it's what most of us think about immediately whenever the topic of energy comes up (and most of us aren't thinking about the city bus either). According to the U.S. Census, nearly nine in ten Americans drive to work; more than three-quarters of us drive alone.[10]

In the past fifty years, the number of passenger cars in the United States has more than doubled. In 1960, the country had nearly 74 million "registered highway vehicles," as the government likes to call them, including some 62 million passenger cars—little deuce coupes and the T-Birds that Daddy threatened to take away, if we judge by the music hits of the day. By 2006, that number was up to over 250 million, including 135 million cars.[11]

Of course, the United States has about 125 million more people now than in 1960, so you would expect the country to have more cars. We're also driving them more than ever. In 1960, we drove an estimated 718 billion miles and needed nearly 61 billion gallons of fuel to do it. By

---

* Using constant 2000 dollars.

† That's "energy consumption per real dollar of gross domestic product," if you're an economist.

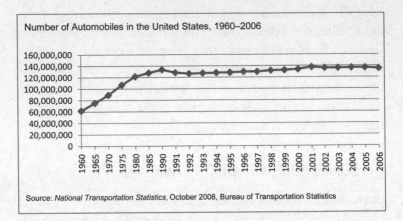

Number of Automobiles in the United States, 1960–2006

Source: *National Transportation Statistics*, October 2008, Bureau of Transportation Statistics

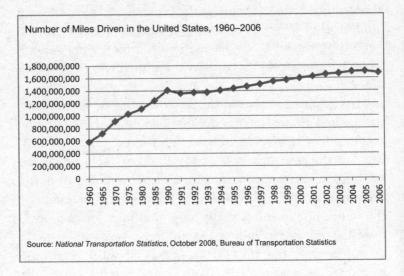

Number of Miles Driven in the United States, 1960–2006

Source: *National Transportation Statistics*, October 2008, Bureau of Transportation Statistics

2005, we drove nearly 3 trillion miles and needed nearly 140 billion gallons of fuel.[12]

In the last few years, Americans have cut back on their driving, particularly during the recession and spike in energy prices that started in 2008. But this is still a big country, and we're still driving way more than before.

For all the attention they attract, cars and trucks don't deserve all the blame. Just look around your home right now. Think of all the stuff that wasn't even invented thirty years ago, let alone cheap enough for the average consumer to buy.

Entire categories of consumer electronics have arrived, and in some cases, such as the VCR, all but departed in that time. Just take a look at how things have changed.[13]

In 1980	By 2001
Government statisticians didn't even ask people whether they had a personal computer	More than half of Americans (56 percent) have at least one computer
37 percent have a dishwasher	More than half (53 percent) have a dishwasher
14 percent have a microwave	86 percent have a microwave
27 percent have central air-conditioning	55 percent have central air-conditioning
43 percent have no air-conditioning at all	Only 27 percent have no air-conditioning at all[14]

But it's not just the stuff we have; we now keep our stuff in bigger houses. The average floor space of a new American house has expanded dramatically in just a generation, increasing from 1,660 square feet in 1973 to 2,521 square feet in 2007,[15] and that's during a period when the average American family was getting smaller.[16] And when people live in bigger spaces, they not only have more space for gizmos, they also have more rooms to light, heat, and cool. If most Americans lived in studio apartments, they probably wouldn't have those three TV sets per home.[17] (Well, some of them might. Some people never want to be more than a few feet away from ESPN.)

This is why we're running in place: While we've become more efficient in some ways, we also continue to use more and more energy.

8. WE IMPORT A LOT OF ENERGY, BUT WE PRODUCE A LOT TOO

Back on *Seinfeld*, on the (many) occasions where George Costanza needed a fake name, he tended to blurt out "Art Vandelay." Mr. Vandelay was a man of many talents, including architect and latex manufacturer, but several times he turned out to be an "importer/exporter." When it comes to energy, the United States is an importer/exporter too.

Almost everyone knows the United States imports large amounts of oil, but it's important to remember that the United States actually produces significant quantities of energy, including oil. We import nearly 60 percent of our oil, but we produce roughly 40 percent, and that's a lot of oil. By any standard, we're a big player in the oil business—the world's number three oil producer, and the number one refiner of crude oil. But we're also the number one consumer, and therein lies the rub, as they used to say in Shakespeare's day. The United States is also:

- The number two producer of natural gas (behind Russia)
- The number two producer of coal (behind China)
- The number one producer of electricity from nuclear power, as well as number one in nuclear capacity
- The number four producer of hydroelectric power from dams, and number two in capacity[18]

But in the end, none of this is enough. Like so many other economic issues, this is simple supply and demand—and our demand continues to outstrip the supply. That's likely to continue. Projecting energy demand is tricky, depending on things that are hard to predict, such as economic growth and new technology. But the consensus is that demand is going to keep rising, and production won't be able to keep up.[19]

Whether the United States could produce more oil, how long it would take, and whether it could be produced without seriously damaging the environment is a major energy debate that we'll come back to later.

9. MOST OF THE OIL THE UNITED STATES IMPORTS DOESN'T COME FROM THE MIDDLE EAST

Nearly two-thirds of Americans believe, wrongly, that we import most of our oil from the Middle East.[20] It's easy to see how people could get that impression. Lingering memories of the 1973 Arab oil embargo and doubts about the "real" reason for the war in Iraq have caused Americans to worry that our foreign policy is being driven by our need for oil or that "we're getting our oil from people who hate us." Or, worse, our oil purchases are funding countries that back terrorists.

Yet while Arab countries in the Middle East are important suppliers for the United States, they are not actually at

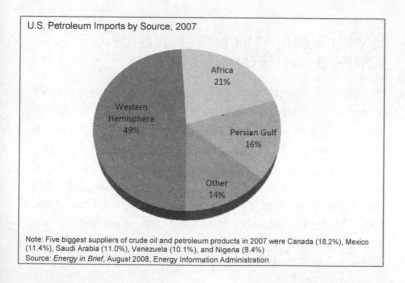

U.S. Petroleum Imports by Source, 2007

Africa 21%

Western Hemisphere 49%

Persian Gulf 16%

Other 14%

Note: Five biggest suppliers of crude oil and petroleum products in 2007 were Canada (18.2%), Mexico (11.4%), Saudi Arabia (11.0%), Venezuela (10.1%), and Nigeria (8.4%)
Source: *Energy in Brief*, August 2008, Energy Information Administration

the top of the list. The top two oil suppliers to the United States are Canada (18.2 percent) and Mexico (11.4 percent); next come Saudi Arabia (11 percent), Venezuela (10.1 percent), and Nigeria at 8.4 percent. Taken together, the Persian Gulf states accounted for 16 percent of our imports in 2006—a sizable amount. But they're not the dominant source.[21]

Sadly, the starring role Middle Eastern countries have played in the ups and downs of gas prices over the years says something about how dependent we are on oil and how even a limited change in the supply can send prices upward here. What's more, the worldwide demand for oil is much higher than it was in the 1970s, so there's even more competition for what is available. Unfortunately, the picture from here on out doesn't look much better. All key experts are predicting that worldwide demand for liquid fuels including oil will continue to grow, and when you take a look at the countries with the largest known reserves of oil, our frenemies in the Middle East are right up at the top.

10. WE'LL NEED MORE OIL IN THE FUTURE—OR SOMETHING THAT WE CAN USE IN ITS PLACE

It's not just that the country's demand for energy is projected to outstrip our supply. We face growing demand for one particular form of energy: liquid fuels that can be used in cars, trucks, and airplanes. Right now, that's mainly petroleum, so you can see the potential problem here. The United States already imports most of our oil, and everyone from President Obama to some eight in ten Americans thinks the country is too dependent on foreign oil.[22] There are other options: liquid fuels such as ethanol, which stretches out petroleum, or liquefied natural gas, which would replace it. Or we can move toward electric cars. None of those options is an easy fix, however.

In fact, one of the frightening things about the government's energy projections is how little is expected to change. The Energy Information Administration projects that use of renewable energy, such as biofuels, wind, and solar, will increase. But even with those increases, based on current trends, the EIA believes fossil fuels will still provide 79 percent of America's energy in 2030.[23]

Projections don't have to come true, of course, and we can all have an Ebenezer Scrooge moment here, asking the Ghost of Energy Future to tell us if these are the things that *must* be or just the things that *might* be. Of course, the answer is always "might be." To change the future, however, we're going to have to start changing the present, and that means taking a serious look at what our options are.

The Great All-in-One Fossil Fuel Centerfold

Anyone who's traveled abroad knows how handy it is to master a couple of basic phrases in the local tongue. If you can say "please," "thank you," and "how much is this garish T-shirt?" you can generally get the help you need. The same principle applies to the energy issue. Master just a few basic phrases and concepts, and you're a long way toward grasping the key points. Here's our fossil fuel centerfold to help you get on top of the energy debate.

Photo: iStockphoto

Type of fuel	How It's Used	Fossil Fuel?	How Much Do We Have?	Renewable?	The Big Question
	Nearly 70 percent of energy in the United States is used for either transportation or electricity.	They're called fossil fuels because they're created when long-dead plants and animals are turned into fuel after lying under the Earth's crust for millions of years. Fossil fuels are responsible for 98 percent of carbon dioxide emissions in the United States. CO_2 is a key cause of global warming.	Do we have enough for current needs? What about the future?	The EPA defines renewables as fuels that are "continuously replenished on the Earth." We don't have to worry about running out of them.	What challenges lie ahead? What are the big decisions the country faces?
Petroleum	Mostly for transportation. Every year, Americans use about a quarter of the oil produced worldwide.	Yes. Whether it's crude oil, diesel, gasoline, or jet fuel, burning it releases carbon dioxide, which adds to global warming.	The United States has about 2.4 percent of the world's known oil reserves	Afraid not.	We have to import it, and it's causing global warming. Plus it won't last forever. The big question is what to replace it with.

(Continued)

Type of fuel	How It's Used	Fossil Fuel?	How Much Do We Have?	Renewable?	The Big Question
Coal	It's used to generate about half of the country's electricity.	Yes, and if you're worried about global warming, it's one of the most dangerous. Burning it releases 25 percent more carbon dioxide than petroleum and twice as much as natural gas.	A lot. The United States has the world's largest known coal reserves, enough to last a century, maybe more.	Nope. Creating it takes millions of years.	Are there ways to burn coal without releasing carbon dioxide? Many experts say it's worth working on, but it won't be cheap. More in Chapter 7.
Natural gas	It generates about a fifth of the country's electricity and is the main source of heat in more than half of U.S. homes.	Yes, but one of the least harmful. Burning it releases less carbon, sulfur, and nitrogen than burning either oil or coal.	The United States has about 3.6 percent of the world's known reserves.	Nope. Just like oil and coal, it takes millions of years to form.	Natural gas can be liquefied and used to replace gasoline. Is that a good idea or not? See Chapters 8 and 12.
Hydroelectric	Dams generate about 6 percent of the country's electricity.	Nope. It's just flowing water, and the process doesn't pollute the air or the water.	A lot, but about half the country's hydroelectric power is in California, Oregon, and Washington state.	As long as the rivers keep flowing, yes.	Hydroelectric power doesn't pollute, but damming rivers can cause problems for fish and wildlife. More in Chapter 10.

Solar	Less than 1 percent of our energy comes from solar. It's used mainly to generate electricity.	Nope. In fact, solar doesn't use "fuel" at all, just the sun.	Plenty, but more in some places (Phoenix) than others (Seattle).	We certainly hope so. No sun, no us.	Solar is in its infancy, but the government hopes that it will be economically competitive by 2015. See Chapter 10.
Wind	It generates less than 1 percent of our power, mainly for electricity.	Nope. Wind is created by weather patterns. There are no plants, animals, or carbon dioxide involved at all.	More than enough, but like solar, some regions (such as the Great Plains) are better suited to this than others.	Yes; the answer, my friends, is blowing in the wind.	We'll need to build a lot more turbines and a grid system to transport the power from the plains to the city folk. See Chapter 10.
Geothermal	Currently it supplies less than 1 percent of our electricity, but California has more than thirty plants up and running	Nope, and the process releases minimal amounts of carbon dioxide.	There's quite a lot in the West, where there are a lot of volcanoes and geysers.	Yes; as long as the Earth's core is hot, we're in business.	Now we use geothermal power that's near the Earth's surface, but some experts say it's possible to tap heat deep inside too. More in Chapter 10.

(Continued)

Type of fuel	How It's Used	Fossil Fuel?	How Much Do We Have?	Renewable?	The Big Question
Biomass and biofuels	Now it's roughly 3.6 percent of our overall energy use. You can burn this stuff in your fireplace, but we also use biomass to generate electricity, and fuels such as ethanol to power cars.	No. We're talking about the current crop of plants and plant waste: trees, corn, even corn husks, peanut shells, and some garbage. Burning it releases carbon dioxide, but since plants absorb carbon dioxide during photosynthesis, it's basically a wash.	It's a big country with plenty of plants and trees and garbage galore.	Yes; unless we wipe them out, the plants will always be with us.	This is a big category, and some forms are more promising than others. More in Chapter 12.
Nuclear	It generates about 19 percent of the country's electricity.	No. Uranium doesn't come from fossils, and generating electricity from nuclear power produces almost no greenhouse gases—about as much as wind power.	The United States has about 6 percent of the world's known recoverable uranium.	Uranium ore is not renewable, but most experts believe there's enough worldwide to go around.	Can we keep nuclear plants safe from terrorists and store the waste securely? More in Chapter 9.

Sources: BP Statistical Review of World Energy, 2009, Energy Information Administration, the Environmental Protection Agency, and World Nuclear Association

CHAPTER 6

DOUBLE, DOUBLE, OIL, AND TROUBLE

Oil is like a wild animal. Whoever captures it, has it.
—*J. Paul Getty (1892–1976)*

Oceans of words have been written about oil, but if you have to sum up the black stuff's influence on the world, it would come down to one thing: oil makes things move.

By that we mean that oil is what we use to move people and things around. Almost all the oil used in both the United States and the world at large is used for transportation. And the ability to travel has remade the world.

It's easy to take traveling for granted these days, especially given the hassles of overbooked or canceled flights and bumper-to-bumper traffic. But let's put that in perspective. In the early nineteenth century, you basically had two choices: a horse or a sailing ship. The vast majority of people never traveled more than 100 miles from where they were born. It took weeks to cross the Atlantic, and four months to go from the East to California by wagon train. There was also a pretty good chance of dying before you got there.

That started to change with coal. Steamships and rail-roads cut transatlantic and transcontinental travel times to a week or less. Now with oil, you can fly around the world in less than twenty-four hours, and the biggest risks will be airport food and lost luggage. Even if you're one of the handful of drivers who obey the speed limit, you can still cover 100 miles in less time than it takes to watch *Indiana Jones and the Kingdom of the Crystal Skull*.[1]

Oil made automobiles and airplanes feasible. It's the reason we can have global markets. American kids can play with toys made in China. Fresh fruit can come from South America to your supermarket without spoiling. Goods can be made anywhere and sold anywhere—and you can pick up and move anytime you want. In a country that believes in new starts and endless possibilities, you can leave that podunk town your parents and grandparents lived in and move to the big city or, depending on your point of view, get out of the dirty, crowded city and commute from your own patch of green in the suburbs. Petroleum helps make that possible. Forgive the pun, but it oils social mobility.

Oil is also amazingly portable. You can shoot it through pipelines, store it for long periods in tanks, load it on ships and trucks, even pump it from one airplane to another in midair. Oil is efficient—it packs more energy punch than any other fossil fuel, and far more per gallon than any of the alternatives, such as ethanol. Oil works so well that it's knocked all the transportation alternatives out of the box.

The fact that oil fuels transportation is what makes it so fundamental to our economy and our society, as well as those of other countries. It isn't just that the price of oil affects how much you pay to drive around; it's that, because every-thing has to be shipped somewhere at some point, the price of oil affects the price of everything. So the challenge is finding out how to find enough oil to move our stuff—or to find other ways of moving our stuff that don't depend on oil.

WHO'S MAKING THE MONEY HERE?

Chances are you opened this book wondering if you're getting ripped off at the gas pump and asking that ever-present question we all have about oil: *Where is all that money really going?*

It'd be nice to have a one-word answer, but even *supercalifragilisticexpialidocious* doesn't do it. To answer that question, we have to talk about the oil business. It's complicated and you'll need to absorb some background info. Put in the time, though, and you can impress the heck out of your friends the next time someone is whining about why gas costs so much. Here's what you need to know.

You know, we know, everybody knows the United States imports most of its oil—about 60 percent in 2007, or about 12.3 million barrels a day. And as we pointed out in the last chapter, about half of that imported oil comes from other countries in the Western Hemisphere: Canada, Mexico, and other places in Latin America.

But what about that 40 percent of domestic oil? The top five oil-producing states are Texas, Alaska, California, Louisiana, and Oklahoma. Fully one-quarter of our "domestic" oil is in fact coming from offshore rigs in the Gulf of Mexico.[2] Domestic production amounted to roughly 5.1 million barrels a day in 2007.[3]

That's a lot by any standard. The United States pumps out nearly 8 percent of all the crude oil in the world, making it the third-largest oil-producing nation behind Russia (12.4 percent) and Saudi Arabia (12.3 percent).[4] That means we pump more oil here than most of the countries you usually think of as oil-rich, including Iran, Venezuela, and Kuwait.

However, if you step back and look at this by regions, you get a picture that matches up with the conventional wisdom. In 2007, all of North America (the United States, Canada, and Mexico) accounted for 16.5 percent of global oil production—a very respectable figure, and a bigger share than either Africa or Latin America. But it's dwarfed by the 31 percent produced by all the Middle Eastern countries lumped together.[5]

So let's take a quick look at some questions that are perplexing Americans.

CAN WE FIND MORE OIL IN THE UNITED STATES, AND HOW MUCH MORE?

Regrettably, the long-term trends don't favor the United States. In the oil business, it's not just how much you pump; it's how much you have *left* to pump. And on both counts the United States is losing ground.

The fact is, U.S. crude oil production peaked in 1970. Ever since then, the trend has been basically downward, largely because North American oil fields are generally "mature." We've been drawing down our oil fields since the 1860s, and it shows. The average U.S. oil well is less productive than it was in 1972. Back then, the average well produced 18.6 barrels a day; by 2007 that had fallen to 10.2 barrels, the lowest level since the government started keeping track in the 1950s.[6] Even the new fields opened up by the Trans-Alaskan pipeline in the late 1970s haven't changed this overall trend.[7]

But the real kicker is who has the reserves—in other words, who's still got oil in the ground. The United States

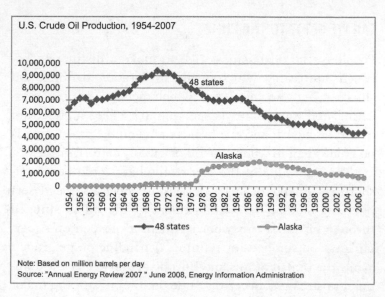

U.S. Crude Oil Production, 1954-2007

Note: Based on million barrels per day
Source: "Annual Energy Review 2007 " June 2008, Energy Information Administration

has about 29.4 billion barrels of "proved reserves," according to BP's *Statistical Review of World Energy*—about 2.4 percent of the world total.* When you lump Canada and Mexico in, North America still has only 5.6 percent of known reserves, or about 69 billion barrels.

The Middle East, by contrast, has *61 percent of all proved reserves*, a staggering 755 billion barrels. Saudi Arabia alone has 21 percent. And most other regions of the world outpace North America—Russia alone has 6.4 percent of all reserves. There's certainly oil in the United States that can be tapped, but overall, we're just not sitting on deep resources in North America.

---

* According to the BP Statistical Review of World Energy, "proved reserves" means oil that "geological and engineering information indicates with reasonable certainty can be recovered in the future under existing economic and operating conditions."

CAN WE GET IT TO THE PUMP?

People in the oil business like to make a distinction between "upstream," which means exploration, drilling, and shipping, and "downstream," which includes refining, marketing, and your corner gas station. It's a long supply train by any standard, and the brand name on the gas station is no guide at all to where the oil originally came from.

The beginning of the upstream chain could be an offshore rig in Indonesia, a derrick in Siberia, or a desert pumping station in Saudi Arabia. Crude oil is shipped through pipelines to seaports, where it's loaded on supertankers and shipped for refining, depending on its grade.* While the United States and Europe are no longer the dominant upstream sources of crude oil, they're still the major downstream refiners who turn it into actual gasoline and other products. The United States is the world's largest refiner of oil by far. We may pump only 8 percent of the world's crude, but we refine more than one-fifth of the world's petroleum products. Almost all the fuel in the United States is refined here, mostly at facilities on the Gulf Coast, although some is refined in Europe and then brought here by ship. Once the oil is refined into products, it's shipped by additional pipeline, tankers, and trucks to factories, gas stations, and airports around the country.

Curious about just where the gas in your tank comes from? Don't even bother to try to figure it out, according to

---

* Crude oil grades are a complex series of ratings that sometimes sound suspiciously like premium coffees (Arabian light sweet crude) or public schools (West Texas Intermediate). They're classified as "light" or "heavy" based on their density, and "sweet" or "sour" depending on their sulfur content. Light sweet crude is the best, because it can be refined into a broad range of products.

the Energy Information Administration. There's so much buying and selling of oil to start with—companies don't just rely on their own wells; they frequently buy crude from each other to get through slack periods or supply problems. And once crude gets to the refinery, it all gets mixed up and processed anyway. The gas in your tank could be from a dozen different countries, or just one or two.

Choke Points

One other key point about this long oil supply chain: there are many things that can go wrong. The system is full of choke points, where the nation's or even the world's oil can get backed up. Some of these are the things that keep people in the State Department awake at night. For example, 20 percent of the world's oil travels on tankers through the Strait of Hormuz in the Persian Gulf, a region bristling with extremist groups and heavily armed states that hate each other.[8] During the Iran-Iraq War in the 1980s, the U.S. Navy ended up escorting tankers through the strait to keep them out of the fighting. This is part of the reason why the United States maintains a strong naval presence there now. The Strait of Malacca in Indonesia, the Suez Canal, and the Panama Canal are also seagoing choke points that could be a problem.

Other things are what keep oil executives up at night. Tankers can run aground, spilling their valuable cargo and ravaging coastal areas and marine life. Pipelines break (a break in an Arizona pipeline pushed up gas prices by 45 cents a gallon locally in 2003); refineries and oil rigs get knocked out by hurricanes (as with Katrina in 2005). Most experts agree that the nation's petroleum infrastructure, just like its roads and bridges, is getting cramped and old. We use pipelines to move 60 percent of all petroleum products inside the country, and the 166,000 miles of domestic pipelines we have are going to have trouble keeping up with demand, particularly in key states such as California, Colorado, Nevada, and Arizona.[9]

ARE WE TAKING ADVANTAGE OF EVERY DROP? AND SHOULD WE?

Remember this exchange from *The Simpsons*?

> **Kent Brockman:** Kent Brockman at the Action News desk. A massive tanker has run aground on the central coastline, spilling millions of gallons of oil on Baby Seal Beach.
> **Lisa:** Oh no!
> **Homer:** It'll be okay, honey. There's lots more oil where that came from.

Homer Simpson isn't the only one who thinks there's plenty of oil out there. A great many people will tell you that there's a lot more oil; we've just put it off-limits with environmental and other rules. To some extent that's true. If you're willing to accept the trade-offs, we can get some more oil in North America.

But even leaving aside the environmental issues—and we'll talk about those in a moment—there's not much "easy" oil left, and there's no fast oil, and there may be no cheap oil either. To squeeze more oil out of U.S. territory takes us into some of the toughest locations for oil exploration on the planet. We're going to have to go to the Arctic, and we're going to have to go to "deepwater" wells more than 1,000 feet under the sea floor. As the Energy Information Administration puts it, "New oil reservoir discoveries are likely to be smaller, more remote (e.g., Alaska), and increasingly costly to exploit."[10]

We've mentioned this already in this book, but it's worth saying again: nothing happens quickly in the energy business. Large oil projects can sometimes take fifteen years or more from discovery to actual production (although it can be less if you're fortunate). Exploration, drilling, refining, and delivering oil all require time and money. For example:

- Alaskan North Slope fields take years to develop in the harsh conditions. The last two major fields developed, Bodoni and Alpine, took eight and six years, respectively, to actually produce oil.[11]
- Offshore rigs aren't much easier, and it isn't uncommon to take a decade to go from discovery to drilling. The Thunder Horse platform in the Gulf of Mexico produces more oil than all of Louisiana, but also cost $4 billion to build and more than eight years to get to production.[12] Canada's Hibernia platform cost $5 billion and took nineteen years to get into production.[13]
- New technology in the oil and gas market takes about sixteen years to go from concept to wide adoption.[14]

Despite the image of the "wildcatter" who takes big gambles seeking out oil in unproven territory, the oil industry is actually quite conservative in betting on new developments. Drilling and coming up empty is a much more serious setback in the wilds of Alaska or on an offshore oil rig. You've spent a ton of money to get there, and you could seriously damage your company if you're wrong.[15]

What's more, a large proportion of these chancier ventures aren't viable unless oil prices are high and likely to continue to be high. From the industry's point of view, the investment simply doesn't pay off unless there's some real money to be made. On the other hand, if the technology

The Trans-Alaskan Pipeline opened up new territory for oil exploration, but even Alaskan fields haven't changed the long-term decline in domestic oil production. Photo by Bruce Green. Courtesy of the U. S. Department of Energy, National Renewable Energy Laboratory

used for exploration, drilling, and extraction advances very quickly, the results could be better. The Energy Information Administration's scenarios turn out much differently depending on how technological innovation affects oil production.[16]

So here's a menu for some of the main choices for increasing U.S. oil production. You can decide for yourself whether they're worth it.

North to Alaska: The Arctic National Wildlife Refuge

There may be some longer-running battles between the oil industry and environmentalists, but not many of them. Both sides seem to approach this debate as if it's the Alamo of energy policy: fight to the bitter end and take no prisoners. They just disagree on who's inside the fort and who's not. Everyone's got an opinion, but here are the basics:

- **Estimated available oil:** Anywhere from 5.7 billion to 16 billion barrels of "technically recoverable" oil (the mean estimate, which is most frequently used, is 10.6 billion barrels).[17] Remember, the United States used about 7.5 billion barrels of oil in 2007.

- **Estimated time to get the fields in production:** Eight to twelve years after Congress passes legislation, presuming no lengthy legal battles, with production reaching its peak ten years after that. So if ANWR was opened in 2010, it would start producing in 2020 and peak in 2030, and gradually decline after that.[18]

- **Estimated impact on prices:** Would reduce the price of a barrel of crude by anywhere between 44 cents and $1.44 in 2028.[19] Just as a reminder, the price for crude oil has ranged from around $30 per barrel to over $130 per barrel in just the last few years.[20]

- **Estimated environmental damage:** Environmentalists say ANWR is one of the last undisturbed Arctic ecosystems on the planet, home to polar bears, caribou, migratory birds, and whales, many of which are protected by international treaties as well as U.S. laws. Since the Arctic summer is short, this is a major breeding and feeding area for species that migrate or go into hibernation during the winter. Advocates for expanded drilling say the development footprint is comparatively small and the necessary roads and pipelines could be routed away from the most sensitive areas, but congressional studies have warned the impact on wildlife could be significant.[21]

Run Silent, Run Deep: More Offshore Drilling

Back in 1990, the federal government put a moratorium on drilling on most of the so-called outer continental shelf in response to concerns from coastal states worried that

Offshore Oil

Outer Continental Shelf (OCS) Areas	Crude Oil (billion barrels)	Natural Gas (trillion cubic feet)
Available for leasing and development		
Eastern Gulf of Mexico	2.27	10.14
Central Gulf of Mexico	22.67	113.61
Western Gulf of Mexico	15.98	86.62
Total Available	40.92	210.37
Unavailable for leasing and development		
Washington-Oregon	0.4	2.28
Northern California	2.08	3.58
Central California	2.31	2.41
Southern California	5.58	9.75
Eastern Gulf of Mexico	3.98	22.16
Atlantic	3.82	36.99
Total Unavailable	18.17	77.17
Total Lower 48 OCS	59.09	287.54

Source: Annual Energy Outlook Analyses, *January 2003, Information Administration*

an oil spill would threaten their shoreline. The trade-off here is exactly the same as it was then: more oil versus the risk of oil spills.

When the price of crude soared in 2008, the pressure to swing back to drilling became intense, with the 2008 election-year chant of "drill, baby, drill" just one expression of it. Still, as we've said before, there aren't many fast solutions in the energy field, and it would take at least five years before you'd see production from new offshore fields.

- **Estimated available oil:** There are about 59 billion barrels total in the outer continental shelf, according to the Energy Information Administration. But 41 billion barrels are in areas that are already open to drilling; about 18 billion barrels are in areas currently off-limits. Almost all of it is in the Gulf of Mexico.[22]
- **Estimated time to get into production:** The EIA figures that if the restrictions were lifted in 2012, production would start in 2017.
- **Estimated impact on prices:** When the EIA looked at this in 2007, the agency said expanded offshore drilling "would not have a significant impact on domestic

crude oil and natural gas production or prices before 2030."

- **Estimated environmental damage:** An oil spill would have significant consequences on the affected shoreline, both for wildlife and for the tourist industry. A spill is not inevitable, of course, but it could be extremely damaging if it happens. Offshore drilling advocates make several important points: (1) advances in drilling technology and contingency measures have made this kind of drilling less risky; (2) the industry has drilled safely in the North Sea and elsewhere; and (3) spills from drilling platforms are fairly rare, compared to other sources.*

Tar-nation: Canadian Oil Sands

So far we've been talking about what people in the industry call "conventional oil"—the crude that the world's been running on for a century. But there's also "unconventional oil," and the higher prices go, the more viable it is.

The hot ticket in unconventional oil right now is the so-called Canadian oil sands, which caused an old-fashioned oil rush in Alberta in 2008. Oil sands are full of bitumen, a form of oil so thick that it's more like tar (which is why these are sometimes called "tar sands"). But you can turn it into crude oil, either by digging up the sand and processing it or by pumping steam or hot water down into the ground to force the bitumen up so it can be refined.

There's nothing new about oil sands—oil producers have known about them for a century and worked on extracting

---

* Statistically speaking, spills from pipeline breaks and leaking storage facilities are the most common, but tend to be smaller; spills from tankers and drilling platforms are rare but involve bigger spills. See the Global Marine Oil Pollution Information Gateway, UN Environmental Program, http://oils.gpa.unep.org/facts/oilspills.htm.

oil from them off and on over the years. But oil sands have two huge advantages over other alternatives to conventional oil. One is that they're available right now. Unlike some of the other alternatives, the technology is proven and ready to go (Canada has been refining oil from oil sands since 1967). And second, this represents a substantial oil source in the hands of one of our closest allies. If you count both oil sands and conventional oil, Canada becomes second only to Saudi Arabia in oil reserves. We probably don't have to worry about Canada turning off the tap, or becoming so chaotic that revolutionaries are routinely blowing up production facilities.

- **Estimated available oil:** If we're looking at what can be recovered right now, that's 175 billion barrels in Canada and perhaps 11 billion barrels in the United States (mostly in Utah and Texas).
- **Estimated time to get into production:** Already there. Canada has been producing oil from these sands since 1967.
- **Estimated impact on prices:** That's the catch. Crude from oil sands costs about 30 percent more to produce than conventional oil, so this process makes sense economically only when prices are high. When oil is cheap, nobody is much interested in pouring money into oil sands. In fact, the crash in oil prices in the fall of 2008 took much of the steam out of the oil sands boom.[23] When oil is commanding high prices, it's worth someone's while to go after this. All in all, oil sands might help us meet our oil needs and reduce our reliance on oil from countries that are unstable or don't like us much, but it won't help much with keeping gas prices low.
- **Estimated environmental damage:** Getting the bitumen out of the ground takes huge amounts of hot water, and

a lot of energy to heat that water. All that heating power—mostly from natural gas—produces massive quantities of greenhouse gases, not to mention the drain on water resources. Controlling greenhouse emissions from oil sands production is a major challenge in Canada.[24]

OKAY, BUT WHY DOES OIL COST SO MUCH?

Yes, we did say we'd answer that question, but we also said there may not be a single answer. And one of the first concepts that we need to understand is that no one actually sets the price of oil. It's the result of a system of bidding and buying and selling that's something like the stock market (and you know how much control we have over that).

In Monty Python's *Life of Brian*, there's a scene where Brian is being chased by Roman soldiers and needs to grab a fake beard right away from a Middle Eastern market (it's a long story). But the shopkeepers are deeply offended because Brian doesn't want to do any bargaining; he just wants to pay the asking price and go. "This bloke won't haggle!" cries the merchant. "Won't haggle?" the other merchants say, shocked.

Haggling over prices has more to do with the energy business—especially the oil business—than you'd think. That's because oil is one big world market—lots of people are selling and almost everybody's buying, and there's plenty of haggling thrown in as well. Oil is one of the truly global commodities, and it's sold on the commodities markets. If you're like most people outside the financial world, your impression of the commodities market is a big trading floor full of people shouting about pork bellies.

That's a good start, because the idea of commodities futures started out with farm products (the New York Mercantile Exchange, the first and still one of the biggest places for oil trading, started out as the Butter and Cheese Exchange). Basically, the system works like this: Let's say you know you are going to want 100,000 barrels of crude oil next January. (We won't ask what you want it for.) On the commodities exchange, you can buy those barrels from a producer in advance (that's why it's called a "future") at a fixed price, say $100 a barrel. The advantage is that you and the oil producer have a deal you can count on at a fixed price. You don't have to worry about not having the oil, and the producer doesn't have to worry about not being able to unload it.

But it's also a gamble. You, the buyer, are gambling that the price won't drop between now and January. If it falls to $90 a barrel, you just lost out on cheaper oil and took a $1 million bath. But the producer is gambling that the price won't go up to $110 a barrel—if it does, the producer is the one who takes the $1 million bath.

So it's not some neat, orderly process that determines the price of a barrel of oil. It's the result of hope, fear, bravado, and panic. Oil prices go up based on what traders think will happen in the future; often that's based on something they heard today that suggests either that there might be a good supply of it available in the months to come or, conversely, that something's afoot that could mean there'll be less of it to buy. From there on out, the laws of supply and demand begin to kick in.

A ROOMFUL OF SUSPECTS

People are always looking for the one guilty party for oil prices, as if this were an Agatha Christie mystery and it was just a matter of Hercule Poirot using his little gray cells to point out the real murderer. But if today's world oil market were a murder mystery, it would be *Murder on the Orient*

Express, where, in the end, everybody on the train did it together.* You'll hear charges and countercharges about why prices are going up and who's getting the money, and it often boils down to "it's not me, it's him." There's some truth in all the finger pointing. In each case, it's *partially* him.

So if the little Belgian detective got all the suspects in a room, you'd be looking at the following:

Growing demand. Every expert, every agency, every study in the field agrees: global demand for oil is up, and it's going to keep going up as China, India, and other developing countries reach the point where they're going to want all the good things oil can provide. This goes for both crude oil and refined gasoline. In the short term, demand depends on numerous things, most notably the overall state of the world economy. During recessions, as we've pointed out before, demand goes down because people produce and ship fewer products and travel less on their own. But even though the world economy and individual countries do go through recessions and downturns repeatedly, the overall economic trend is upward, and that means more demand for oil.

Tight refinery capacity. Crude oil is no good to anybody until it's refined into something else. As we said, the United States is still the world's biggest refiner of oil, by a long shot. And for most of the past twenty-five years, there's been excess refining capacity worldwide. But now the market has tightened up considerably and refineries are running at close to capacity. There haven't been any new refineries built in the United States since the mid-1970s, although there's been a lot of expansion of existing ones. The oil companies complain that it's hard to build more because local

---

* Sorry if we ruined the ending for you, but hey, the book's been out for seventy-five years.

residents often object to living near an oil refinery.[25] The expense is a factor too—what the National Petroleum Council calls an average-sized refinery can cost up to $3 billion to build.

Differing regulations. Then there's trying to fill what basically amounts to a series of special orders. Some states have passed mandates requiring different fuel blends or ethanol to meet clean-air regulations. That may be good for the environment, but it complicates the refinery business. They have to produce different kinds of blends, which both adds to their costs and means they can't always shift gasoline around the country to meet the demand. If one state requires all gasoline to have 10 percent ethanol in it, for example, then companies can't just ship gas in from another state, which doesn't have such a requirement, if there's a shortage.

Low inventories. The oil companies don't keep as much crude oil and gasoline stashed away waiting to be used as they used to. Partly this is due to declining domestic production, partly due to the ability to buy crude on spot markets, and partly due to their "just-in-time" production strategies, which are designed to reduce costs. In the early 1980s, U.S. oil companies had enough gasoline on hand to cover forty days of use; by 2006 that had fallen to twenty.[26] Some experts also believe that the Strategic Petroleum Reserve run by the government allows the industry to reduce inventories.* Normally, this is fine—buying oil whenever you want it on the commodities markets is usually cheaper

---

* The government established the Strategic Petroleum Reserve after the Arab oil embargo in the early 1970s to insulate the United States against future oil shocks. In theory, it's supposed to provide at least 90 days' worth of net imports. Where is it? It is sitting in a series of caverns in Texas and Louisiana and contains more than 700 million barrels of crude oil. You can find out more about it online at www. whoturnedoutthelights.org.

than keeping it in storage. But if anything goes wrong—if refineries go offline, or pipelines break, or the international markets go crazy—there's not much cushion to fall back on. Hurricanes Katrina and Rita proved that in 2005. They knocked out refineries, pipelines, and oil rigs along the Gulf Coast, and drove up gas prices nationwide.

WHO'S AFRAID OF BIG BAD OIL?

Americans don't trust oil companies, and they didn't even before they saw Daniel Day-Lewis beat a rival to death with a bowling pin in *There Will Be Blood*. When a Gallup poll asked about impressions of various business sectors in August 2008, the oil and gas industry ranked right down at the bottom, with just 16 percent of respondents having a positive impression of the industry. Some other industries, such as real estate, had more dramatic falls, but the oil companies were consistently low.[27]

When most Americans look around for people to blame for high gas prices, they look to the oil companies (and oil-producing countries) first. Nearly six in ten blamed these two for high gas prices in 2008, according to Gallup, although there was more than enough blame to go around (half blamed speculators, while 46 percent blamed increased demand).[28]

There's a very simple and undeniable fact underlying this: when you're suffering from high gas prices, the oil companies are making money. Lots of it. In 2007, the oil industry overall had revenues of $1.9 trillion, with three-quarters of that going to the five "integrated" companies that work both upstream and down (ExxonMobil, Royal Dutch Shell, BP, Chevron, and ConocoPhillips). That added up to profits of $155 billion.[29] And that was before the sharp rise in gas prices in 2008. But, as experts point out, the profits were "upstream," in drilling and pumping crude, and not "downstream," at the refineries and the gas pump.

So what, you might say. Big Oil is Big Oil. But in fact, Big Oil isn't as big as it used to be. At least, the companies we think of as Big Oil aren't as big as they used to be. There's been a real power shift in the world oil industry in the last couple of decades. Once upon a time, the big oil companies such as ExxonMobil, BP, and Royal Dutch Shell owned this business. They found it, they drilled it, they shipped it, and eventually one of their employees pumped it into your car and cleaned your windshield while they were at it. Oh, sure, a large proportion of the oil was overseas— but only the oil companies had the technology to find it, the market clout to sell it, and the backing of powerful Western countries if there was any trouble. If you were an oil-producing Third World country, the safest course was to grant an oil concession to one of the major companies and just cash the checks.

That's changed. You've heard of Shell and ExxonMobil, but have you ever heard of Aramco? PDVSA? Pemex? These are the state-owned companies that control the oil in Saudi Arabia, Venezuela, and Mexico, respectively—and they're not alone. One of the biggest trends of the last couple of decades has been the rise of the national oil company— essentially, government-owned corporations in the oil business.

And not surprisingly, these companies are in the countries that have the most oil. Oil accounts for 40 percent of Saudi Arabia's entire economy and more than 70 percent of its tax revenues.[30] Other Middle Eastern countries are much the same. For most of these countries, oil is way too important to be left to a foreign company.

Since most of the world's oil reserves are outside North America, that means they're also mainly in the hands of national oil companies. State-owned companies own more than 80 percent of the proved reserves in the world. The five major private oil companies control only 3.6 percent.[31]

Again, you may say, so what? It's their oil; surely these countries can handle it any way they want. But is this big shift in oil power bad for us? Maybe, in some cases. The private oil companies basically have one clear-cut agenda—to make money. We may not always approve of the way they go about it, but we understand it. But a national oil company is only partly a company—it's also a government agency. Market forces are only one of the things they're going to factor into their decision making. For these companies, oil isn't just a business; it's a major national asset, one that's deployed in their national interest.

Some are investing in order to ensure their own oil supply down the line. The Chinese national company, for example, is investing heavily in Africa for this reason. That's good for the Chinese and good for the African countries that benefit from the investment. Other countries, such as India, use their national oil companies to keep prices as low as possible at the pump.

But there could be a darker side. Right now, the open market for oil works pretty well for everybody. But it's not the only way of selling oil. Maybe a national oil company would be more likely to set up side deals with preferred customers, namely, that country's friends and allies. Or maybe they would use their control of oil to weaken and harm the United States for political reasons.

CAN THE UNITED STATES BREAK ITS ADDICTION TO OIL?

Despite all the talk about alternative fuels, fuel efficiency, conservation, and new kinds of cars than run on electricity or hydrogen fuels, we can't just snap our fingers and be done with oil. We use it in too many ways in every nook and cranny of our economy.

President Bush famously said the United States was addicted to oil, and the country does display some of the symptoms of addiction—including repeatedly trying to cut back on oil consumption, only to relapse to our old habits. But going cold turkey is not really an option. Maybe it's better to think about oil as a "maintenance medication." We might like to stop relying on the meds, but we just can't do it until we're on the road to recovery.

Peak Oil, or When Will We Run Out?

Given that we depend on oil even to move the oil itself from place to place, the big question is: Are we running out?

The short answer is yes, but nobody knows exactly when. As we've mentioned before, there's general agreement that energy demand is going to increase pretty dramatically and may well outstrip production. But this is a different question. There's only so much oil out there—everyone agrees on that. And since we're not making any more dinosaurs, we're not actually creating any more oil. So at some point we're going to run out.

There's a whole movement out there based on chewing over the issue of "peak oil," which is the exact point at which production maxes out and begins to decline.* After oil hits its peak, there is less and less of it to pump out of the ground. When you've eaten half a chocolate bar, you've still got half left, but there's less and less each time you take a bite. You get the picture. The term comes from "Hubbert's Peak," named after geologist M. King Hubbert, who in the 1950s correctly predicted U.S. oil production would peak in 1970.

There's no question that the world hitting peak oil in the near future, when most of us will still be alive, is a real possibility. But peak oil discussions often have a *Mad Max*–style end-of-the-world tone to them, as if once the peak was passed, the immediate result would be a social collapse, complete with postapocalyptic biker gangs in hockey masks and feathered

---

* There's even an Association for the Study of Peak Oil, which you can find more about at www.peakoil.net.

headdresses roaming the highways seeking whatever fuel was left. Peak oil is not the same as no oil. But it *is* kind of a point of no return. And if world oil production were to begin to decline year after year, then the consequences could be severe. Prices would shoot up dramatically—picture the oil shocks of the 1970s, but on a permanent basis, causing a global recession that could extend for a very long time.[32] It would all depend on when and how fast.

Estimates of peak oil are all over the map. When the GAO examined the various projections in 2007, all they could conclude was most analysts believed it would occur sometime before 2040. But some thought it had already happened, and others thought it was seventy-five years off or more.[33] Major energy organizations, such as the International Energy Agency, predict a slow increase in oil production through 2030, although not enough to keep up with demand.

Oil predictions are also famously hit-or-miss. The U.S. Bureau of Mines once predicted that U.S. oil production would reach its maximum in two to five years—back in 1919.[34] Other experts have predicted global peak oil in 1989, 1994, 2000, and 2003.[35] The problem is that experts don't agree on exactly how much oil is out there, how much of it we can actually get, and what we should be willing to do to squeeze out every last drop.

For one thing, there are a lot of people with an interest in being cagey about this. When it comes to the oil-estimating game, there's a slew of players, all with different interests. Oil is spread out all over the world, and there are countless incentives not to be honest about it. Some governments and oil companies put out pretty honest data, and others don't. The best estimates are that there's about 13 trillion to 15 trillion barrels of oil out there worldwide—but only about one-third of it can be pumped with existing technology, according to the National Petroleum Council.[36]

Yet it's technological innovation that makes peak oil a moving target. Britain's North Sea offshore drilling consistently produced more oil than anticipated over the 1980s and 1990s, as companies found new ways to squeeze more oil out of the fields. Even though U.S. oil production peaked in 1970, the subsequent decline has been pretty slow, partly because of new ways of making older wells produce (such as injecting steam or carbon dioxide into wells to push more oil out). There are also questions about whether we should look and drill for oil every place it could possibly be. In the United States we've been arguing for years about whether more federal land (such as the Arctic National Wildlife Refuge) should be opened for oil exploration and drilling.

Another key factor here is price. As oil becomes scarcer, it becomes more expensive, and when it becomes more expensive, it becomes financially worth it to chase oil in more difficult places (such as deepwater offshore areas, where the oil is more than 1,000 feet down). It also becomes more viable to go after so-called unconventional oil, such as oil sands and shale (see pages 93–95). Higher prices also prompt people to cut back on their oil use, stretching the supply. That won't mean oil becomes cheap again, but it does mean that you can get it if you're willing to pay for it.

So reaching peak oil need not be a crash—it could be a more gradual decline. After all, once U.S. oil production peaked in 1970, it showed a fairly gentle overall decline over decades, partly thanks to new discoveries in Alaska and offshore. But a gradual decline merely means we have more time to find alternatives to oil, whatever they may be. And, of course, if we've got alternatives to oil coming online—if biofuels, hydrogen cells, or other kinds of vehicle power are available—then that softens the blow. If not, then we could be in for some bad times and very expensive gas.

The Return of Snidely Whiplash:
Are Evil Speculators to Blame?

Speculators. Even the word has a villainous air to it. A 2009 opinion survey by Public Agenda showed that nearly seven in ten Americans believe speculators are responsible for high oil and gas prices,[37] and they're not the only ones. When oil prices slipped the surly bonds of Earth in mid-2008, some elected officials called on regulators to investigate whether "speculators" were artificially driving up prices.[38] Most experts think surging worldwide demand for oil was the chief culprit, but others have questioned whether 2008's wild price swings were just too radical to be the normal ups and downs of the market.[39] The idea of speculators toying with oil and gas prices is bad enough that it's worth taking a moment to understand exactly what the charge is.

First of all, we should be clear about what it's not. The fact that oil prices whoosh around doesn't, by itself, prove that speculation is to blame. Since oil is traded on open commodities markets, prices are inherently twitchy (much like the stock market). Traders will bid low if they think oil will be plentiful and cheaper down the road, and they'll bid high if they think it will be scarcer and cost more—and of course, they don't actually know.

It doesn't always work out the way even very, very smart people think it will. Southwest Airlines acquired the right to over 70 percent of the jet fuel it thought it needed for 2008 through a "fuel hedge fund," based on oil prices of roughly $50 a barrel.[40] During the summer of 2008, the airline executives looked like geniuses because oil was selling for nearly three times that much. But at year's end, with prices near $30 a barrel, Southwest was hurting.[41] They bet on oil prices, and at first they won; later they lost.

Obviously, most players in oil futures are drillers, refiners, and companies such as Southwest that actually want to use the oil for something. But there are also lots of people in the oil futures market who aren't really in the oil business at all; they just want to make money by buying and selling it. And for many, that's the definition of speculation right there—people who make money by trading and betting on oil prices going up and down.[42] The difference between a "speculator" and an "investor" is that you envision one twirling a Snidely Whiplash handlebar mustache and the other in a gray three-piece suit.

How much speculation has been going on in the oil markets of late? In 2008, federal regulators initially concluded that speculation was not a major factor in the wild price ride.[43] Several months later, however, regulators determined that Vitol, a Swiss company, was trading for its own profit, and at one point held 11 percent of all the oil contracts on the New York Mercantile Exchange.[44] By 2009, the Commodity Futures Trading Commission had reversed course and announced it would consider new limits on speculative trading.

Vitol (which sounds like an energy drink, not a trading company) didn't make the headlines, but Enron did.[45] Enron traders took advantage of a badly constructed California energy deregulation plan to essentially transfer California's electricity out of state and then sell it back into the state at an inflated price, making nearly $500 million and causing needless blackouts in the process.[46] Recorded phone calls show Enron traders chortling over "Grandma Millie," who didn't understand why her power went out.[47]

The company's reckless and criminal management led to

its collapse in 2002, and two Enron traders pleaded guilty in the California case, just part of the Enron scandal. But Enron's case goes beyond just playing the markets. Enron traders and executives broke the law.

It's not so easy to draw a bright line between investing, speculation, and manipulation of the market. Since much oil trading takes place outside the United States, tracking down and stopping speculators is even more complex. Not everyone thinks having more "outsiders" in energy trading markets is such a bad thing.[48] Outsiders can bring more money into the system and perhaps spur more discoveries and quicker recovery of oil.

In the end, maybe the best way to reduce speculation in the oil markets would be to reduce oil's power in our lives. If we weren't so dependent on it, we wouldn't be so vulnerable to its price swings, whether they're exaggerated by speculators or not.

CHAPTER 7

YOU LOAD SIXTEEN TONS AND WHAT DO YOU GET?

I was born one mornin' when the sun didn't shine,
I picked up a shovel and walked to the mine.

—*Merle Travis*

Coal. It's the best of fuels and the worst of fuels. For anyone worried about U.S. dependence on imported energy, coal is a tantalizing option. The United States has coal mines in twenty-seven states, with Wyoming, West Virginia, Kentucky, and Pennsylvania leading the way.[1] As a result, this country is often considered the Saudi Arabia of coal because of the vast supplies available here, enough to last well over two hundred years, according to estimates by the Energy Department.[2]* Currently, coal is used to generate about half of the electricity Americans use (nuclear power and natural gas are the other big sources).

But for anyone worried about global warming and climate change, coal is one of the worst choices of all. Burning

---

* Some critics question the estimates, saying that it assumes that you can mine anywhere, such as under major cities, in national parks, and so on. But everyone agrees that we have huge supplies of coal.

coal releases massive quantities of carbon dioxide, at least the way it's done now. Former vice president Al Gore, who won the Nobel Peace Prize for his work on global warming, considers coal so dangerous that he once asked "why there aren't rings of young people blocking bulldozers and preventing them from constructing coal-fired power plants."[3] For Gore, moving away from coal is a top priority for battling climate change, and Gore is hardly alone. Dr. James Hansen, the NASA climatologist who began calling for action on global warming twenty years ago, also wants to halt construction of any new coal-burning power plants unless they have technology that prevents the release of carbon dioxide.[4] As we explain later, it may be possible to develop that technology down the road, but it's not ready yet.

BIG FIGHTS AHEAD

The complication of course is that most Americans are worried about reducing the country's dependence on foreign energy *and* fighting global warming. That means that the decisions we face on coal could lead to some of the most pointed debates the country faces. Over the next decade, you can expect to see major disputes on these and other questions:

- **Should we spend taxpayer money to find reliable ways to get rid of the carbon dioxide produced by burning coal?** Carbon dioxide is a major contributor to climate change, but since we have so much coal available to us in the United States, maybe we should try to clean it up.
- **Should we build any more power plants that use coal?** Should we build them only if they have very expensive new technology to capture the carbon dioxide before it's released into the atmosphere? Some of these decisions

will be at the state and local levels, where local utility companies are regulated.

- **Should we spend taxpayer money to promote the conversion of coal into something like gasoline?** That way we might not be as dependent on imported oil, but there are some real environmental drawbacks.
- **Should we tax the carbon dioxide that comes from burning coal?** Or maybe we set up a system where companies that want to burn coal can buy and sell permits to release a certain amount of it into the air. (Yes, it's a little strange, that second one, but we'll go into it more later.)
- **Or should we decide that it's just too expensive and risky to keep trying to "fix up" coal?** Maybe our money and effort would be better spent on other kinds of energy, such as solar, wind, and even nuclear power. None of these alternatives is perfect, perhaps, but none is nearly as risky as coal when it comes to climate change.

There is a lot to think about, and choices we make now will have repercussions for years. Building a power plant is massively expensive, so plants are generally expected to operate for three or four decades. This is not like trying out a new breakfast cereal, where you can just toss it out if you don't like it, and at worst you're out several bucks. Whether a new power plant uses old coal, clean coal, nuclear power, natural gas, solar panels, or whatever, we'll be living with the consequences for a generation or more.

WHAT WE OWE TO OLD KING COAL

Coal has been a major player in American life since the Civil War. In the 1800s, it powered the railroad and provided gaslight in major cities from New York to San Francisco. It fueled the country's industrial growth in the 1900s. By the 1960s, it was mainly used in power plants to generate electricity, and today, over 90 percent of coal is used this way.[5] Unless you're sitting on a beach reading this (and honestly, if you are, stop reading and enjoy the sun and the surf), you may well be using some coal-generated electricity right now.[6]

At the moment, coal is the cheapest source of energy we have.[7] Energy giant BP, which publishes a respected annual survey on how fuels are produced and consumed worldwide, reports that coal has been the fastest-growing fossil fuel in use for the past five years.[8] There's really not much mystery why—because coal is readily available and affordable compared to other energy sources. Since the 1950s, the United States has been a major coal exporter, with nearly half of what we export going to Europe.[9]

As an energy source, however, coal has had pros and cons for a long time. On the positive side, coal is affordable, reliable, and the United States has plenty of it. Over the years, the industry has made commendable progress in reducing the air pollution and acid rain associated with coal. There is the possibility of developing technology down the road that would capture the carbon dioxide produced when coal is burned so it couldn't contribute to global warming—that is, if we're willing to invest in it. The coal industry provides jobs and economic benefits in communities nationwide, and some regions rely heavily on the jobs and other economic benefits the industry provides.*

---

* See, for example, the American Coalition for Clean Coal Electricity (ACCCE) at www.cleancoalusa.org and Friends of Coal at www. friendsofcoal.org.

On the other hand, mining coal has been a dangerous, dirty, and sometimes fatal business, and it is still considerably riskier than many other jobs.[10] Controversies over strip mining and mountaintop mining have plagued the industry, and environmental groups are nearly unanimous in their opposition to coal for a whole variety of reasons.[11] The Natural Resources Defense Council (NRDC), for example, says that using coal is "among the most destructive activities on earth, threatening our health, fouling our air and water, harming our land, and contributing to global warming."[12]

So what are the major issues behind the battle over coal?

Problem #1: Smog and Acid Rain

Burning coal produces sulfur and nitrogen oxide— chemicals that create smog and acid rain. The Clean Air Act of 1970, later amended, requires power plants to cut emissions of sulfur and nitrogen oxide.[13] Consequently, it's much safer to breathe these days. According to the GAO, "Although some older power plants emit high levels of these substances, significant advancements have been made" in newer power plants.[14] The country's progress in fighting smog and acid rain is impressive, but that doesn't mean the country is making headway on climate change. As we mentioned before, more than half of Americans mistakenly believe that "by reducing the level of smog in the United States, we've gone a long way to reducing global warming."[15] It's just not the case. To fight climate change, we'll need to reduce carbon dioxide emissions that come from burning fossil fuels, especially coal (see page 115).

Problem #2: Mercury

According to the Environmental Protection Agency, burning coal also produces about 50 tons of mercury every year.[16] Mercury can pollute rivers, streams, and oceans and end up in fish and shellfish, which can be harmful to the humans and animals who eat them. The Bush administration and environmental groups battled in and out of court for years on regulating mercury emissions; in early 2009, the U.S. Court of Appeals opened the door for the Obama administration to formulate new regulations.[17] Don't be surprised if this argument continues for quite a while.

Problem #3: Carbon Dioxide (aka the Whopper)

But coal's whopper problem by far is that burning it releases carbon dioxide, and carbon dioxide is the whopper worry in climate change. In the United States, more than a third of the carbon dioxide released into the atmosphere

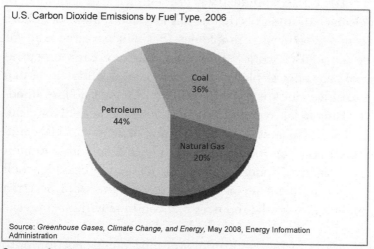

U.S. Carbon Dioxide Emissions by Fuel Type, 2006

Coal
36%

Petroleum
44%

Natural Gas
20%

Source: *Greenhouse Gases, Climate Change, and Energy*, May 2008, Energy Information Administration

Our use of coal produces half of our electricity, and more than a third of our greenhouse gas emissions.

comes from coal. If we look just at emissions from generating electricity, coal accounts for over 80 percent of the problem.[18]

Today, the crux of the debate is about whether scientists and engineers can figure out a reliable, practical, affordable way to burn coal without letting carbon dioxide get into the air, and whether the country should build any more coal-fired utility plants until that new technology is put in place.

Major coal producers and power companies (and many others, including President Obama) say that the idea is worth pursuing, not only for what it can do to reduce emissions here but also for what it can do worldwide—lots of countries have lots of coal and lots of coal-fired electricity plants. In fact, many experts believe that if we don't move ahead to develop the technology here, foreign governments and companies will, and that's already happening to some extent. There's too much coal out there, many experts believe, for us not to try this route.

The U.S. coal industry is promoting the development of "clean coal" and says that it "has invested billions of dollars in new advanced technologies that will pave the way for design of the world's first pollution-free, carbon-neutral coal-based power plant."[19] The American Coalition for Clean Coal Electricity is also investing millions of dollars in advertising to let the public know about the potential for clean coal. (See "Clean Coal: The Ad Wars" on page 125.) Depending on how you look at it, that's either a demonstration of the industry's solid commitment to solving coal's carbon dioxide problem or an attempt to pull the wool over the public's eyes so the industry can continue with business as usual.

Many in the environmental movement question both the goal and the industry's commitment to it. In 2008, for example, Al Gore argued that clean coal "might be possible, many years from now . . . But there's not a single demonstration project in the United States. They're not doing anything . . . to put substance . . . to the slogan, 'clean coal.'"[20]

SO CAN IT BE DONE?

And that brings us to the $64,000 question about coal today (or maybe $64 billion is the better figure). Is it possible to develop a way to burn coal that doesn't release carbon dioxide into the air and doesn't contribute to global warming, and how long would it take? And the answer is . . . You guessed it: It depends on whom you ask. Scientists and engineers are looking at a couple of options, but the one that seems to attract the most attention involves converting coal into a gas, removing the carbon dioxide, and either burying it or storing it or finding some way to use it that doesn't release it into the atmosphere.*

The wonk terminology is "capturing and sequestering," or, as it's sometimes called, "carbon capture and storage" or CCS. There are a handful of plants worldwide using different versions of the process. Norway has four CCS projects, one of them operating since 1996, and has launched an eight-year development program aimed at making the process affordable.[21] Also, a Swedish company called Vattenfall opened a CCS plant in Brandenburg, Germany, in

---

* See National Governors' Association, "Securing a Clean Energy Future—A Governor's Guide to Clean Power Generation and Energy Efficiency," 2008, p. 22, for discussion of several options for removing carbon dioxide from the process of burning coal (www.nga.org/Files/pdf/0807CLEANPOWER.PDF). The National Energy Technology Laboratory also has a good discussion of government research on the storage issue at its website at www.netl.doe.gov/technologies/carbon_seq/

2008. According to the company president, Lars G. Josefsson, "It represents the first ever transition from lab to reality. Our intention is to make a decisive contribution to global climate protection."[22] The U.S. government has invested some money in an experimental capture-and-sequester project called FutureGen. The Energy Department pulled the plug on the project in 2008, but Energy Secretary Steven Chu has expressed interest in reviving it with some modifications.[23] You'll just have to stay tuned on FutureGen.

IF MIT CAN'T DO IT, WHO CAN?

The basic science seems to be less of a problem than the engineering, economic, and political challenges. The UN's Intergovernmental Panel on Climate Change believes the general idea is worth pursuing, [24] as does the Obama administration, which included money and loan guarantees to support carbon capture and sequestration technologies in its budget for 2010.[25] In 2007, a dozen experts from the Massachusetts Institute of Technology (MIT) judged the concept promising enough to call for full-scale demonstration projects, saying these should be mounted right away.[26] But the MIT team also underscored the "enormous system engineering and integration challenge."

When a dozen MIT experts say something is doable, it's probably worth listening to them. On the other hand, when a dozen MIT experts call something an enormous engineering challenge, we tend to think it must indeed be an enormous engineering challenge. The MIT brain trust concluded that extracted carbon dioxide could be stored safely with careful engineering and government oversight, but other scientists aren't as sure. Some have raised questions about whether the carbon dioxide could seep out of its storage places in the ground or seabed and become hazardous.[27]

WHEN IT'S CHEAPER TO EMIT

The much bigger problem, however, is whether the process will ever be economically viable, and what government should or should not do to promote it, or perhaps even require it. In 2008, the Advanced Coal Technology Work Group, a group of industry leaders, environmental groups, and engineers and other experts, called for "a variety of regulatory, financial, and other incentives" to "accelerate" the development of sequestration and related technologies.[28] Howard Herzog, one of the MIT experts, points out that "all major components of a carbon capture and sequestration system are commercially available today." So why isn't it up and running? As Herzog puts it, "It is almost always cheaper to emit to the atmosphere than sequester."[29]

As a result, there are more than six hundred plants and nearly fifteen hundred individual generators in the United States that use coal as the main fuel for electricity.[30] As of 2006, none were capturing and sequestering carbon dioxide. Plus, it's not like the United States can walk away from coal-generated power in a New York minute. You can't just knock down a coal-fired power plant and then put up something that generates energy another way overnight. So we have to decide what to do, given that reality.

A NEW POWER PLANT COMING TO A LOCATION NEAR YOU?

Right now, utility companies, state and local governments, and environmental groups are battling about building new power plants all over the country. Politically speaking, getting a license to build a coal-fired power plant is dicier than it used to be. Coal's backers often argue that new plants keep electricity reliable and affordable and that—even without the still-in-the-womb CCS technology—new coal plants are much more energy efficient, and they disgorge much less nasty pollution-creating and acid-rain-causing stuff than older ones. Some believe that new coal-burning power plants could be retrofitted with the technology when it's ready, though the MIT team thought this would be enormously expensive and maybe not even practical.[31] However, American Electric Power Company is moving ahead with plans to retrofit its coal-burning facilities. Its head, Michael Morris, says that he is "an absolute believer that we need to retrofit the existing fleet while we move to energy efficiency, renewable, new nuclear."[32]

Opponents generally see building a new coal-burning power plant as spending an enormous amount of money on a type of energy that's outmoded and dangerous. Put that money and effort into alternatives, they say, such as solar, wind, nuclear, and others that don't release carbon dioxide and don't cause climate change.

Whatever the pluses and minuses, some experts expect a slowdown in construction of coal-fired power plants until the debate is resolved. Not that many investors, they point out, like to sail into new projects with so much uncertainty in the air.

GOOD-BYE CHEAP COAL?

The other big fight, mainly at the national level, is about plans to tax or charge companies that release carbon dioxide, so that alternatives such as solar, wind, and nuclear power (and other possibilities) could get a leg up. Making it more expensive to release carbon dioxide into the air could also give coal and utility companies a stronger incentive to get the capture-and-sequestration technology up and running.

The downside, of course, is that those higher costs will be passed through the system and eventually show up in customers' utility bills. Since coal generates nearly half the country's electricity, that means an awful lot of us will end up paying more. Americans currently pay less for electricity than consumers in much of Europe where there is more reliance on renewables and nuclear power.[33] Being able to get so much of our electricity from cheap and domestically-mined coal is one reason why.

Taxing carbon emissions is a relatively straightforward idea; the major debatable points are how high the tax should be, whether the money should be devoted to alternative energy or returned to taxpayers to offset higher electricity costs, and whether it would jeopardize economic growth. Cap-and-trade plans (which we explain in more detail in Chapter 15) are a variation on this theme. Essentially, the government would set up a system where companies could buy and sell permits to emit carbon dioxide. In the end, it has the same effect as a carbon tax—it makes it more expensive to release carbon dioxide into the air. There's considerable wrangling among experts about which system would work better and be fairer. Weirdly, the fact that cap and trade is complicated and hard to grasp actually makes it a little more acceptable in the political world. Since a straight-out carbon tax would be simpler for voters to understand, they might get mad at politicians who voted for it.

FROM GASLIGHT TO GAS TANK?

And what about that other big idea for coal? Could you transform coal into a liquid fuel that we could put into our cars instead of gas made from petroleum that we have to import? As a matter of fact, you can, but whether that's a good idea or a horrible one depends on whom you ask.

It may sound like some form of medieval alchemy, but the doability is relatively easy. Liquid coal technology dates back to the 1920s and supplied a large part of Germany's fuel needs during World War II. There's already a South African company named Sasol that produces some 150,000 barrels of fuel a day using coal; South Africa adopted the technology during the apartheid era, when it faced an international oil embargo because of its human rights abuses.[34]

To the head of the world's biggest coal company, Gregory Boyce, CEO of Peabody Energy, here's an obvious answer to America's energy predicament: "We have the resources right under our feet."[35] But the environmental side of the picture is not nearly as pretty. In fact, when it comes to the risk of global warming, using liquid coal for cars is worse for the environment than using plain old gasoline (and using plain old gasoline is bad enough). A study by the Environmental Protection Agency looked at two options. One was making and using liquid coal with the technology we have right now—that is, technology that doesn't capture and sequester carbon dioxide. That would actually double the amount of greenhouse gases our cars spew out—a pretty horrifying thought to anyone worried about global warming.[36]

Even with a carbon capture makeover (and remember, even the MIT techies say that's not going to be ready anytime soon), using liquid coal is still a little worse than using

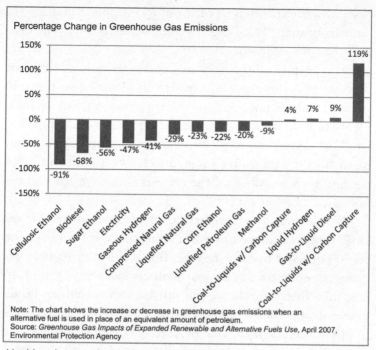

Percentage Change in Greenhouse Gas Emissions

Note: The chart shows the increase or decrease in greenhouse gas emissions when an alternative fuel is used in place of an equivalent amount of petroleum.
Source: *Greenhouse Gas Impacts of Expanded Renewable and Alternative Fuels Use*, April 2007, Environmental Protection Agency

Liquid coal, even using the best technology available, would be more damaging to the environment than gasoline.

gasoline.[37] Meanwhile, some potential alternatives, such as biofuels, battery-powered cars, and liquefied natural gas, would reduce the amount of carbon dioxide we pump into the air if we used them broadly.[38] A spokesperson for the National Environmental Trust, an advocacy group, called the liquid coal idea "the snake oil of energy alternatives."[39]

So here we are with decisions to make and some contradictory information being thrown at us. That's not unusual these days. Think of all the conflicting information you have to wade through to pick out a cell phone plan or get auto insurance. Thinking through what to do about coal isn't so much different.

And despite all the back-and-forth, there is a fair bit of agreement about the basic trade-offs. Using coal the way we use it now is not good for the environment, but we have plenty of it, and we don't have to import it. Clean coal that wouldn't contribute to global warming is a possibility, but it's still in the planning stage, and it won't happen unless we get to work pronto on the engineering, investment, regulatory, and other aspects.

Liquid coal for cars is relatively easy to make.[40] If we close our eyes to the environmental costs, it could reduce our oil imports to some degree. But is this a defensible decision when there are so many other options we could invest in? Where should we put our resources? Which kinds of energy use should we encourage?

Coal was the backbone of America's energy strategy for more than a century, and we all owe a debt to the people who mined it for our benefit. Some of them died; many were injured; many developed black-lung disease, cancer, and other serious illnesses. We owe a debt to the coal industry that brought the United States into the twenty-first century. And cheap electricity makes a huge difference in the day-to-day lives of millions of people. But when we think about the future, coal's role is not as clear. Can it be cleaned up to serve us for another couple of centuries? Or, when we look to the future, are other options better bets?

Clean Coal: The Ad Wars

Most people probably had a better chance of understanding the idea of "clean coal" before the two sides started putting ads on TV. The commercials about clean coal are about as helpful to most of us as commercials about opposing political candidates or about the differences between Macs and PCs. (Yes, Mac users wear jeans and PC users wear suits and ties. We get it.) So let's start off with a reality check on clean coal. Here's what you need to know:

- If you're thinking about the emissions that cause air pollution and acid rain, coal is much cleaner than it used to be, at least here in the United States.
- But if you're thinking about the emissions that cause global warming, coal is still one of the most dangerous substances on the planet.

Experts who spend their careers contemplating things like this say there is technology that would allow coal to be burned without releasing greenhouse gases, but it is expensive, and it also can't happen overnight. The process needs testing, and engineers need to get some of the kinks worked out (like where to put the carbon dioxide once they've removed it). The goal set by the electricity industry is to have 40 percent of the country's power plants using the cleanup technology (generally called "carbon capture and storage" or "carbon capture and sequestration") by 2050.[41] Miley Cyrus will be looking at the big 6-0 by that time, so it's not exactly around the corner.*

---

* She was born in 1992, if you're interested.

Developing clean coal technology is a huge challenge. But figuring out how to use the world's abundant coal supplies without contributing to global warming could also present huge business and economic opportunities, so it's a tough call on how much we should invest in it and count on it.

Just because this is a subtle issue doesn't mean that advocates pro and con aren't trying to give you the answer in thirty seconds or less. Maybe you've seen these ads on TV:

- A grandmotherly woman sits by a lake and tells us, "I believe in the future." An earnest-looking young man, maybe a college student, also believes in the future. Then we hear from lots of different kinds of folks who believe in "protecting the environment," "energy independence," "technology," and "American ingenuity."* The ad is attractive, and it makes clean coal's most persuasive case: we can use coal without harming the environment if we make up our minds to do it and are willing to take on the challenge. Fair enough. You won't be at all surprised to learn that the sponsor, the American Coalition for Clean Coal Electricity, represents energy companies such as Arch Coal, Duke Energy, and Peabody Energy. It also includes an array of electricity companies and cooperatives including American Electric Power, Arkansas Electric Cooperative Corporation, Seminole Electric Cooperative, and Luminant. (You can see the whole list at www.americaspower.org/Who-WeAre/ACCCE-Members.) Many of these electricity providers use many different energy sources, including nuclear, wind, and solar, but they want coal left in the mix.†

---

* You can see this ad and others at www.americaspower.org/News/Ad-Archive.

† Another ad in this series features President Obama talking about the potential of

- Not to be outdone, the Sierra Club, National Wildlife Federation, Natural Resources Defense Council, and other environmental groups crafted a series of responses. In the cleverest, a guy wearing a hard hat (which we assume is supposed to make you look like an engineer) invites TV viewers to take a tour of a "state-of-the-art clean coal facility." He opens the door and walks into an empty wasteland—no buildings, no equipment, a few tumbleweeds, and some rocks. "Amazing!" he gushes. Reminding viewers that coal is a leading cause of global warming, he points to the empty desert in back of him: "But the remarkable clean coal technology you see here changes everything."* That's the point. Clean coal technology doesn't actually exist right now.

Faced with mounting opposition to new coal-fired power plants, the coal industry, which employs over 80,000 people in mining operations alone,[42] sees clean coal as a lifeline, so it's entirely reasonable that they would make their best case for it. And if the "it" is investing in developing the technology for the future, there's not as much controversy as you might think. Al Gore loathes the way coal is used now, but he's open to clean coal as a concept: "If it can be created, if it can be paid for, if it works, then wonderful," Gore recently said.[43] "But let's don't pretend that it exists now. It does not." Of course, some of the ideas popular among environmentalists— ideas like a national "smart" electric grid, tidal power, and cellulosic biofuels—don't actually exist now either.

---

clean coal during his 2008 campaign. He comments on the country's success in getting a man to the Moon and then says: "You can't tell me that we can't figure out how to burn coal that we mine right here in the United States of America and make it work."

* See the entire ad at www.thisisreality.org.

And that's the catch: the likelihood that many Americans will come away from the pro-clean-coal advertising believing that the technology is already available or will be soon. When they hear that coal is cleaner than it used to be (which it is), they may assume it no longer presents any environmental hazards at all (which is definitely not true). And looking at a thirty-second ad that highlights the sizzle, not the steak, hardly any of us would grasp the magnitude of the challenge or the money involved.

On the other hand, many Americans may also come away from the anti-clean coal ads thinking that the idea is mere fantasy that doesn't warrant even a second of their attention. In reality, clean coal deserves a hearing in the court of public opinion even if this country never builds another coal-fired power plant. Others elsewhere will, because coal is just too cheap and plentiful worldwide. So if there's a chance to counteract the environmental dangers, maybe it's in our interest to work on making truly clean coal a reality. After all, climate change won't just affect countries still using coal the old-fashioned way, somehow leaving the United States magically unscathed.

In many respects, moms probably have the best advice for anyone watching the clean coal ad wars. Talk is cheap, moms are fond of saying, and even at TV advertising prices, talking about cleaning up coal is easier and cheaper than actually doing it.

CHAPTER 8

IT'S ALL RIGHT NOW (IN FACT, IT'S A GAS)

I am a big fan of natural gas, and I don't know why it's not considered sexier.

—*Jay Leno*[1]

With all due respect to Jay Leno, natural gas has always been around, and it's probably never been sexy. If you don't believe us, you might want to check out the Roman historian Plutarch. Ancient Rome produced some of the best stories of sex and violence before premium cable, and Plutarch wrote about some of it. He also may have been one of the first people ever to write about natural gas. Many experts believe the oracles and "eternal fires" Plutarch described were actually natural gas that had seeped above-ground and caught fire from lightning.[2] A nice historical footnote, don't you think?

We won't claim natural gas can predict the future (and the spotty track record of the ancient oracles is a good argument against that). But whether it's sexy or not, natural gas does have some attractive qualities as an energy source.

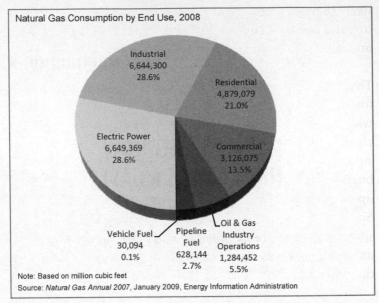

Natural Gas Consumption by End Use, 2008

Industrial
6,644,300
28.6%

Residential
4,879,079
21.0%

Electric Power
6,649,369
28.6%

Commercial
3,126,075
13.5%

Vehicle Fuel
30,094
0.1%

Pipeline
Fuel
628,144
2.7%

Oil & Gas
Industry
Operations
1,284,452
5.5%

Note: Based on million cubic feet

Source: *Natural Gas Annual 2007*, January 2009, Energy Information Administration

Natural gas is an enormously flexible fuel. It's used for heating houses, to generate electricity, and in industry. Should we use it more for our cars?

Currently, natural gas supplies almost a quarter of the United States' overall energy.[3] It's a flexible fuel that can generate electricity, heat houses, and light up stoves, or—in a compressed or liquefied form—replace gasoline or diesel in cars, buses, trucks, taxis, and the like. Natural gas is also used in industrial processes such as glass melting and waste treatment.[4] (And to think Plutarch and his pals were practically worshipping the stuff.) Right now, more than half of American homes are heated with natural gas.[5]

The Department of Energy is projecting that natural gas will be the fuel of choice for more than half of the country's new electricity generation plants over the next twenty years, or to put it another way, utilities are almost twice as likely to build a new plant that uses natural gas as they are to build one that runs on coal or renewables such as wind and solar.*

So why is natural gas such an energy superstar lately? There are several reasons, but a major one is that of all the fossil fuels, it's the least harmful to the environment by a wide margin. According to the government, it "burns more cleanly than other fossil fuels. It has fewer emissions of sulfur, carbon, and nitrogen than coal or oil, and when it is burned, it leaves almost no ash particles."[6] Burning it emits up to 60 percent less carbon dioxide than burning coal.[7] Since millions of Americans rely on natural gas, it is responsible for about 20 percent of the nation's total carbon dioxide emissions. That's something, certainly, but it is less than oil and coal.[8]

If you were a superstar yourself in chemistry class, you may know that natural gas is mainly methane.[9] Methane is also a potent greenhouse gas, although it's generally considered a secondary threat: Carbon dioxide is the King Kong of global warming.[†] There are multiple sources of methane problems in the United States. Energy production is one, but so are decaying landfills and "enteric fermentation," which basically refers to the digestive processes of

---

* The Department of Energy's Energy Information Administration projects that 53 percent of new plants will use natural gas, 18 percent will use coal, 22 percent will use a renewable source such as wind or solar, and 5 percent will use nuclear power. Energy Information Administration, *Annual Energy Outlook 2009*, p. 72, www.eia.doe.gov/oiaf/aeo/pdf/trend_3.pdf.

† Methane is a potent greenhouse gas, but it is less than 9 percent of overall U.S. greenhouse gas emissions. In contrast, carbon dioxide makes up over 82 percent. EIA, "Emissions of Greenhouse Gases Report," www.eia.doe.gov/oiaf/1605/ggrpt.

cows and other livestock.[10] We'll stop while we're ahead on the cattle digestion digression. If you want more details, the EPA's website has them in its section on greenhouse gases and ruminant livestock.* (And you thought your tax dollars weren't paying for anything fun!)

Stacked up against many of the fuels we have available now, natural gas looks pretty good. Let's say you're running the local power company and you know your region is going to need more electricity than you can produce right now. What goes through your mind? You've heard the federal government may be adding fees or increasing taxes based on carbon emissions. You don't want consumers and their local elected officials frothing at the mouth about electricity rates. You'd like to get something built without hordes of people in the community protesting and filing lawsuits.

Suddenly, natural gas begins to shine. It releases fewer carbon emissions than coal. It doesn't get local residents as riled up as a nuclear power plant might.

Natural gas also has considerable pluses as an alternative transportation fuel. The environmental advantages are persuasive enough for leading mass transit systems to turn to it more and more as a fuel for city buses. In 1996, 95 percent of the country's public buses used diesel; by 2007, nearly 14 percent of public buses were running at least partially on some form of compressed or liquefied natural gas.[11]

So with all that natural gas has going for it, what's the problem? Well, there are questions about how much natural gas the country has, what we're willing to do to get it,

---

* If you want to ruminate over ruminant livestock, go to www .epa.gov/rlep/faq.html.

how much we'll need in the future, and how much it will cost if we begin using more and more of it. Unless you skipped over a lot of pages getting here, you know that supply and price go together like love and marriage, a horse and carriage, and so on. (Yes, Frank Sinatra if you're older; *Married with Children* if you're younger.)

So let's start with the basics.

HOW MUCH NATURAL GAS DOES THE UNITED STATES HAVE, AND IS IT ENOUGH FOR THE FUTURE?

If you're looking for a crisp, definitive answer here, forget it. Figuring out how much natural gas the United States has and will have down the line is not as easy as it sounds. Projections can vary depending on several factors. Improved exploration and drilling techniques can make it possible to get to natural gas that wasn't easily available before, meaning that we would have more of it. Rising prices for natural gas and a good economy can encourage more investment and more production, upping the supply. On the other hand, there are disputes about where energy companies should be allowed to drill for natural gas and what kinds of techniques they should be allowed to use. Finally, there's the questions of how much we're going to need and what we plan to use it for. Not surprisingly, some experts believe the United States has abundant supplies; just a few years ago, other experts were painting scenarios that are eerily similar to the mess we're now in with oil.

According to the Department of Energy, the most recent estimates suggest that the United States has nearly 238 trillion cubic feet of natural gas.[12] That may sound like a lot, but it's a small fraction of the world supply. As the Federal Energy Regulatory Commission puts it, "About 96 percent of the world's proven natural gas reserves are outside of North America. At the same time, the United States is consuming about 25 percent of the world's annual natural gas production."[13]

You can imagine how this has the potential to get out of hand. We're using more natural gas than we have in our own country, and we're importing the rest. Right now, the United States mainly imports natural gas from Canada, and the numbers aren't astronomical yet, certainly not compared to our oil imports.[14] But looking down the road is a little less reassuring.

According to the Department of Energy, the worldwide supply of natural gas should be enough to meet demand over the next several decades.[15] However, we can't assume that we're the only ones looking at the natural gas option. Everyone else in the world is looking for a cleaner fossil fuel, and if they increasingly choose natural gas too, that could mean more competition and rising prices.

Would you like to guess which countries have the most natural gas reserves in the world? Russia, Iran, and Qatar.[16] (At least Qatar seems to like us.) So could natural gas become a foreign policy problem just like oil? Some experts are already worrying about how to handle Russia, which has the world's largest natural gas reserves. Russia has made noises about starting some sort of OPEC of natural gas with itself and its national company, Gazprom, as the kingpin.[17] Gazprom may sound like a high school dance with blaring heavy metal music to you, but it's now a major player in the world's energy equation. You'll hear more about it in years to come.

What about the natural gas we have here? The government projects that supplies will increase in the next twenty years mainly because better drilling techniques allow energy companies to remove it from unconventional sources such as shale formations.[18] But the government experts also caution that some of their projections are far from certain.

Others are more optimistic. The Potential Gas Committee at the Colorado School of Mines calculates that the United States natural gas reserves are about a third higher than government estimates.[19] The American Clean Skies Foundation, an industry group, points out that there are twenty-two shale basins in the United States that could, according to their calculations, supply enough natural gas to last a century.[20]

A century is a good long time, so that sounds somewhat reassuring. But Chris Nelder, a journalist who specializes in covering energy issues, points out that the Clean Skies projections are based on current consumption levels. And of course, the whole point of getting excited about natural gas hinges on using more of it. Nelder writes, "If we intend to shift electricity generation toward natural gas and away from coal, then we can already toss out [that] assumption."[21] Plus there's the idea of using liquefied natural gas to replace gasoline.

WHAT SHOULD WE DO TO GET IT?

Not only are there disputes about how much natural gas there is, there are disputes, similar to those in the oil industry, about how far we should go to get it. For example, one of the breakthroughs that make it possible to get more natural gas from more places more efficiently is hydraulic fracturing, or "fracking." It uses water, pressure, and sand to enlarge crevices belowground so it is easier to remove the natural gas. (The process is also used for oil.)[22] Natural gas industry leaders point out that the new technologies mean producers can supply more natural gas with fewer wells, less waste, a smaller footprint, and less use of explosives. They also say they work hard to protect aquifers and groundwater when they use the fracking process.[23]

But many environmentalists remain unconvinced. They say the process leaves millions of gallons of polluted water underground that could seep into fresh water and drinking water sources.[24] The Environmental Protection Agency has determined that the process poses little threat to underground sources of drinking water,[25] but the EPA's conclusions were controversial, and disputes about the process continue.[26] There are also environmental concerns about the pipelines needed to transport natural gas from here to there.

TO DRILL OR NOT TO DRILL

The bitterest disputes, however, seem to center on where to allow drilling and where to prohibit it. As a society, we've decided to close off some areas of the country to oil and natural gas exploration and mining by designating them as protected wilderness, wetlands, or parks. This is why we have bans on drilling in the Great Lakes and legislation such as the Endangered Species Act.[27] Restrictions such as these, taken together, mean that about 250 trillion cubic feet of natural gas cannot be tapped—an amount that would essentially double the current U.S. reserves.[28] Some people see these restrictions as shortsighted and foolish. Others see them as a blessing.

Senator James Inhofe (R-Okla.) believes that many of those who oppose expanded drilling are both seriously misinformed and operating from groupthink. He complains that there are "those who are simply opposed to drilling anywhere, anytime and will go to all lengths to prevent it from occurring." For Senator Inhofe, the "tactics employed to stop exploration and production of new natural gas sources under the pretense of 'environmental protection' are costing this country dearly."[29]

On the other side of the fence, groups including the National Wildlife Federation and the Natural Resources Defense Council are working to bar more drilling on federally protected lands.[30] They point out that an area such as Desolation Canyon in Utah could be open to energy development if drilling advocates have their way. Desolation Canyon was designated as a protected wilderness by the U.S. Bureau of Land Management, which called it a "place where a visitor can experience true solitude—where the forces of nature continue to shape the colorful, rugged landscape."[31] For drilling opponents, the idea of "condemning these lands to a future of oil rigs and gas pipelines," as one critic put it, is a sacrilege.[32]

Most Americans don't follow the details of the debate over protecting the environment versus looking more broadly for domestic energy sources including natural gas. And to be fair, the same tensions are beginning to emerge regarding safeguarding wilderness and wildlife versus putting solar panels and wind farms in prime areas.

Public opinion polls don't shed much light on Americans' genuine preferences. In summer 2008, with gasoline skirting the $4 mark, a number of polls showed rising public support for drilling for oil and natural gas offshore and in other protected areas.[33] At the same time, the vast majority of Americans voice strong support for the concept of protecting species, habitat, and the beauty of the land itself: 77 percent of Americans say they worry about "the loss of natural habitat for wildlife," with 44 percent saying they worry about it a great deal; 68 percent say they worry about the "extinction of plant and animal species," with 37 percent saying they worry about it a great deal.[34]

IS NATURAL GAS THE ANSWER TO OUR PROBLEMS?

Natural gas has some genuine and undeniable benefits, but there are also some big ifs.

- **Increasing imports?** Using more natural gas, especially for transportation, could help reduce the country's dependence of foreign oil, but even now, the domestic natural gas supply is not keeping up with consumption, and consumption would increase if we start using it more for our cars. So unless we increase domestic supplies, we'll have to import more and more.

- **Unacceptable drilling risks?** Using it for electricity, for heating, and as a fuel for cars and trucks is much less harmful to the environment than using coal or oil, but do the new exploration and extraction techniques for natural gas pose unacceptable risks and trade-offs? Maybe it's time for a full-scale review of the evidence from some knowledgeable, disinterested parties such as MIT, the GAO, and others.

- **Will the investment be there?** There's another big if that many Americans will probably find frustrating. It takes a ton of money to find, extract, and transport natural gas, and it will take even more money to build new natural gas power plants, switch coal-fired plants to natural gas, and set up a system for broad use of natural gas as a fuel for cars and trucks. (You didn't really think all of that was going to be cheap, did you?) Unfortunately, the question of whether energy entrepreneurs and investors are willing to fork over the money depends in part on the price of oil. When oil is expensive, there's more incentive to invest in natural gas because it becomes a more attractive and competitive alternative. When oil prices drop, the motivation tends to slump too.

- **How long will it last?** There is another energy fact of life we need to keep in mind. For all its appeal and environmental advantages, natural gas is a fossil fuel that won't last forever. "Every time you use it, you have to wait another billion years for something to create some methane,"[35] is how natural gas entrepreneur Rodney Waller of Range Resources characterizes the dilemma. Many environmentalists just don't see the point of investing tons of money and effort into switching from one fossil fuel to another. From their perspective, it seems better to move full speed ahead to wind, solar, and other renewables without the natural gas detour.

TICKET TO RIDE

At the same time, there is more than a little interest in using natural gas as our ticket out of the waning age of fossil fuels. After all, we have natural gas; we know how to work with it. Right now, at least, it's a more practical and affordable option than some of the longer-timeline alternatives such as wind and solar. T. Boone Pickens specifically promotes using liquefied natural gas in cars as a transition fuel while the country moves over to renewable energy.

But if the United States decides to rely more on natural

gas, if we're going to use it as a "bridge fuel," the question remains: a bridge to what? We'll need to invest in renewable energy so it will be ready when the time comes, and we'll need to plan for a future without as much natural gas even as we make the decision to use it more now.

CHAPTER 9

TIME FOR THE NUCLEAR OPTION?

And Lord, we are especially thankful for nuclear
power, the cleanest, safest energy source there is,
except for solar, which is just a pipe dream.

—*Homer Simpson*

The country's first nuclear power plant opened the year moviegoers were headed out to see *The Bridge on the River Kwai*, and Jerry Lee Lewis was topping the record charts (and scandalizing parents and teachers) with his rendition of "Whole Lotta Shakin' Going On."* Five decades later, there are more than one hundred nuclear reactors scattered over thirty-one states.[1] Nuclear power is a significant source of electricity in some states, including Vermont (where it generates 70 percent of the state's electricity), Connecticut, New Hampshire, Illinois, and South Carolina (where it produces about half).[2] Apart from running U.S. Navy submarines and ships, nuclear power is used almost exclusively to generate electricity, and overall, it supplies

---

* For those of you who are detail-oriented, it was 1957 (www.fifties web.com/57tunes2.htm).

There are more than one hundred nuclear reactors generating electricity in the United States today. Source: Nuclear Regulatory Commission

about a fifth of the country's needs.[3] So the question for this chapter is: Should we use it more?

The very idea of nuclear power seems to produce mixed feelings in the United States. Most of us have a nuclear plant in our own state, and there's not much evidence that vast numbers of us are walking around in a perpetual state of fear about it. Polls suggest that the country is divided on the issue: 47 percent of us say we'd be willing to have a nuclear power plant built in our area as a way to address the country's energy problems, but 47 percent of us say we wouldn't.[4]

Partly it's history. Feelings about nuclear power were certainly upbeat in the 1950s, when ad campaigns provided cheerful drawings of families zipping to the supermarket in their nuclear-fueled wood-paneled station wagons (even as the government was encouraging people to dig bomb shelters in the backyard).* However, most Americans of a certain age can remember the hair-raising accident at Three Mile Island in Pennsylvania in 1979 and the much more harrowing disaster at Chernobyl in Ukraine in 1986 (see pages 156–161 for more on exactly what happened in both places). Those accidents (and maybe the film *The China Syndrome*, featuring considerably younger versions of Jane Fonda and Michael Douglas) have shaped public thinking to some extent. Even before these events, however, the nuclear industry was coping with sluggish investment, opposition from environmentalists, and resistance from communities where new reactors were planned.[5] From 1973 through 2008, not a single new nuclear power plant was licensed in the United States.[6] The industry was basically in pause mode.

Lately, with the United States facing so many questions about its reliance on fossil fuels, nuclear power looks like it might become an energy Comeback Kid. According to the Nuclear Regulatory Commission, the government agency that licenses nuclear reactors, about two dozen new applications have been submitted since 2007.[7] During his 2008 presidential bid, Senator John McCain proposed building forty-five new nuclear power plants by 2030;[8] President Obama—who has conceded the downsides of nuclear power—also says that it is "unlikely we can meet our aggressive climate goals if we eliminate nuclear power as an option."[9] Even some environmental activists, including Greenpeace cofounder Patrick Moore, endorse more nuclear energy—although many remain opposed.[10]

---

* Check out the acclaimed documentary *The Atomic Café* on this sometime.

THE EMISSIONS ARE MISSING

Looking at the energy sources the United States currently uses, nuclear power is the hands-down winner when it comes to generating electricity without adding to global warming. Unlike oil, coal, and natural gas, using nuclear energy for electricity produces virtually no greenhouse gases.[11]

It's also the only "emissions-free" option that's a significant source of energy right now, producing one-fifth of our electricity. Looked at another way, more than 70 percent of the country's emissions-free electricity is coming from nuclear power, compared to 22 percent for hydropower and less than 5 percent for wind, solar, and geothermal combined.[12]

Nuclear energy also has one big advantage over some

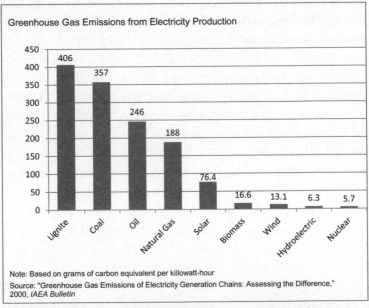

When it comes to generating electricity, nuclear power produces fewer greenhouse emissions than just about any other fuel we have now.

equally environmentally friendly alternatives such as wind and solar (again, as they currently operate). With nuclear energy you don't need to wait for the wind to blow or the sun to shine to start pushing the kilowatts out the door. It basically works nearly all the time. This may seem like an obscure point to you, but if you're running a utility company and your customers expect electricity anytime they want it, it's one you can't afford to ignore.

As we mentioned in our pop quiz earlier, some greenhouse gases are released during uranium mining and other associated activities, but given humankind's current dependence on fossil fuels, it's almost impossible to get anything done without emitting some greenhouse gas. Human beings expel greenhouse gases when we breathe, so we need to keep a little perspective here. According to experts at the UN's International Atomic Energy Agency, the entire nuclear power process—soup to nuts—emits about the same amount of carbon dioxide per hour of electricity produced as wind power or hydropower.[13] That's about as good as it gets given the existing state of the technology. It's important to let that sink in a moment, because surveys show that more than half of Americans believe that nuclear power contributes at least a little to global warming, and one in five of us thinks it contributes a significant amount.[14] Nope, it doesn't.

CHEAP THRILLS

Nuclear power has one other potential advantage: it's relatively cheap once the plant is built and is up and running.[15] In most cases, coal is the cheapest way to generate electricity, but at least as it's currently used, coal is a major culprit on the global warming front. Natural gas is somewhere in the middle, and there have been differing assessments about where its costs are headed.[16] Right now, wind and solar power are substantially more expensive. Their costs should fall in the future if the country works on expanding and developing them, but we're not talking next week or next year; we're talking up to a decade.

Ah, but don't ignore that little phrase, "once the plant is up and running." The process of building nuclear power plants and going through the various approval processes needed to get online costs a ton of dough up front. Nuclear reactors are big and complicated, and their construction involves a lot of concrete and steel, so it's not like throwing up some prefab toolshed in the backyard. Building one can run from $6 billion to $8 billion, according to the Nuclear Energy Institute, an industry group,[17] although other estimates are somewhat higher.[18]

What's more, the industry has an eye-popping history of cost overruns. When the Congressional Budget Office looked at construction records from 1968 to the present, it found that on average, nuclear power plants ended up costing at least double what was projected.[19] Then there are the costs associated with getting a license to operate and various approvals at the national, state, and local levels. When environmental and/or community groups take their concerns about a nuclear power project to court, there are legal costs and delays as well.

Most of us wouldn't want getting a license to build a nuclear reactor to be too easy, but all that vetting and contingency planning does make nuclear energy more expensive. When you include the cost of building and licensing, and giving investors a return on their money (this is capitalism, after all), nuclear power loses all of its cost advantage over natural gas and coal.[20]

Not surprisingly, finding ways to make building plants more affordable (through regulatory changes, loan guarantees, utility rate adjustments, and other approaches) is a major goal of nuclear power advocates. A recent report from the Congressional Budget Office concluded that, given its start-up costs, nuclear energy will remain more expensive than coal and natural gas unless the country establishes some kind of tax or financial penalty for releasing carbon dioxide into the atmosphere.[21] If there were taxes or permit fees for carbon dioxide emissions, utilities that rely on coal and natural gas would have to pay it, and utilities that relied on nuclear (or wind, solar, or hydropower) wouldn't. A team of MIT experts also concluded that government regulations promoting energy with very low carbon emissions would "give nuclear power a cost advantage," but the group also underscored the need to reduce the construction costs and delays that make nuclear power a fairly risky financial proposition for potential investors.[22]

FLORIDA GOES NUCLEAR

The state of Florida is currently working to expand its reliance on nuclear energy, and the decisions there suggest how these considerations work out in real life. Florida Power and Light is planning two new reactors for its Turkey Point facility, near Biscayne Bay.[23] Florida officials are so eager to have them built that they're allowing the utility to charge customers a little over $2 more per month to cover the pre-construction costs.[24] When the new reactors come online in a dozen or so years, Florida customers will start saving about $1 billion annually (remember, nuclear power is actually cheaper once all the pieces are in place).[25] According to the company, the new plants will provide enough electricity for 725,000 homes while avoiding releasing some 280 million tons of carbon dioxide into the atmosphere. This is the equivalent, FPL says, of taking more than a million cars off the road every year.[26] At least that's the plan.

So why didn't Florida Power and Light go with wind or

solar instead? Well, in fact the company is making some smaller investments in both.[27] Despite the advantages of wind power, it hasn't gotten much traction in Florida. To take advantage of winds blowing from offshore, the turbines would typically need to be built near the beach. Many consider rows of windmills a majestic and awe-inspiring sight, but Florida's beachfront hotel owners don't seem to be among them.

What about solar? Florida is the Sunshine State, after all. The problem here is partly scale and the limits of current solar technology. According to experts, it would take nearly 90 square miles of solar panels to generate as much electricity as the two new Turkey Point reactors.[28] What's more, the reactors will operate 90 percent of time, while the solar generators only operate about 25 percent of the time.[29] As James Tulenko, the director of the University of Florida's Nuclear Lab, puts it: "Utilities are going to have real trouble meeting their energy needs without nuclear, which keeps working day and night. It doesn't depend on the wind blowing or the sun shining."[30]

TEN THOUSAND YEARS OF RADIOACTIVITY

Okay, things seem to be going just swimmingly for nuclear power in Florida, so why would anyone have any doubts? Let's walk through some of the major worries one by one.

What About the Leftover Radioactive Waste?

Nuclear power doesn't release greenhouse gases, but it does leave behind a grab bag of radioactive leftovers that need to be managed and stored. Unfortunately, this is not something you can just throw into the Dumpster. While some of it is not especially dangerous and can be handled safely after a few weeks, some of it is so hazardous that it needs to be

isolated for ten thousand to one million years.[31] Having to
store something very, very carefully for ten thousand to one
million years is a fairly mind-blowing responsibility. What
do you store it in? Where do you bury it so it's still safe if
there's an earthquake or some other catastrophic event? Do
you put up signs so humans ten thousand years from now
will know to avoid the place? Ten thousand years is a long,
long, long time. Just to put that in perspective, ten thousand years ago human beings lived in nomadic tribes that
were still learning to domesticate animals.[32]

Unlucky Yucca

Right now, each nuclear plant stores its own waste ("spent
fuel" is the term specialists often use), and, as we said up
front, there are nuclear power plants operating all across
the country: so far at least not too much has gone wrong
on this front.* However, given how bad it would be if something did go wrong, experts have weighed all sorts of ideas
for storing nuclear waste including putting it in polar ice
caps, burying it beneath the ocean, shooting it into outer
space, leaving it where it is, and "placing it deep underground in a geologic repository."[33]

The recommendation from the National Research Council (a group of experts who advise government on scientific
issues) was that "geologic disposal remains the only scien-
tifically and technically credible long-term solution."[34]

---

* According to the NRC's "Fact Sheet on Storage of Spent Nuclear
Fuel" (April 2005), there are basically two ways of doing it. It's either
kept in "fuel pools" with "very thick steel-reinforced concrete walls"
or in "casks" made of metal or concrete designed to "resist situations
such as floods, tornados, projectiles, and temperature extremes"
(www.nrc.gov/reading-rm/doc-collections/fact-sheets/storage-spent-
fuel-fs.html).

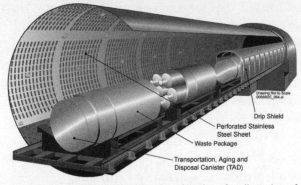

Drip Shield
Perforated Stainless
Steel Sheet
Waste Package
Transportation, Aging and
Disposal Canister (TAD)

Government experts looked at several options for disposing of
nuclear waste—including sending it into space—but they
concluded that burying it deep within the Earth in a specially
designed container like this one would be safest. U.S.
Department of Energy, Office of Civilian Waste Management

Armed with the scientists' admonition to bury the stuff,
the Energy Department proposed a single storage facility
for all of the nation's accumulated nuclear waste at Yucca
Mountain, Nevada. Get it all together, put it into one specially designed, supersafe place, and watch it like a hawk
was the general idea. Unfortunately, Yucca Mountain turned
out to be a "best-laid plans gone astray" story, with delays,
cost overruns, and wide-ranging criticism of the plan.[35] In
2009, the Obama administration proposed terminating additional funding for Yucca Mountain, even though it acknowledged "that nuclear power is—and will remain—an important
source of electricity for many years to come and that how the
nation deals with the dangerous byproduct of nuclear reactors is a critical question that has yet to be resolved."[36]

Does anybody have any better ideas? Experts at MIT
who studied the future of nuclear power several years ago
called radioactive waste disposal "one of the most intractable problems facing the nuclear power industry throughout
the world."[37] In 2009, when they updated their analysis, they

concluded that "the progress on high-level waste disposal has not been positive."[38] In the end, however, the MIT brain trust concluded that the advantages of nuclear power outweigh the risks. Secretary of Energy Steven Chu has promised to "begin a thoughtful dialogue on a better solution for our nuclear waste storage needs."[39]

The French, by the way, who rely heavily on nuclear power, take a different approach to nuclear waste. Learn more about how the French use nuclear power (and what they do with the nuclear waste) at the website at www.who turnedoutthelights.org.

Suppose There's Another Accident?

Although accidents like Three Mile Island and Chernobyl unleashed broad and continuing discussion about the safety of nuclear power, it is important to remember that there are over 400 nuclear reactors operating worldwide, and there's really only been one major accident (Chernobyl) causing significant loss of life.[40]

Surprisingly (or not so surprisingly, if you take a bleak view of human nature), much of the danger seems to lie with the people who do the work, not the design of the plant itself. But that still leaves plenty to worry about: human error, laziness, stupidity, and that timeless standby, "Gee, I didn't think it would be a problem."

How do you make sure the human beings working with nuclear power take all needed precautions, follow all the rules, and check and double-check to make sure nothing has slipped by? The Nuclear Regulatory Commission is the principal watchdog. In the United States, nuclear reactors are licensed and inspected by the NRC, which keeps resident inspectors at each plant.[41] NRC also sends out engineers and specialists to conduct anywhere from ten to twenty-five inspections at each site each year.[42] Since the thought of having a federal government inspector residing where we work would probably lead most of us to go bald pulling our hair out at the roots, the plan looks reasonably demanding and meticulous on paper.

Napping at the Reactor

Unfortunately, there have been some unnerving slipups. A couple of years ago, a guard at Pennsylvania's Peach Bottom plant was so alarmed that his colleagues were dozing off during work that he wrote to the NRC. When there was no response for several months, a videotape of the catnappers found its way onto the evening news.[43] As a mortified NRC official admitted, the agency was "not successful in uncovering the problem" and was reviewing its "processes to see what improvements can be made."[44] Well, what else could he say?

A few years earlier, supervisors at Ohio's Davis-Besse nuclear plant managed to hide their discovery that corrosion had eaten a hole in a container unit even though the plant had complied with multiple NRC filings. How did they get away with it? They lied on the reports.[45] It's precisely these kinds of errors that make people wonder just how safe the nation's nuclear plants are.

The GAO reviewed the NRC's performance in 2006 at the request of Congress and determined that the agency was improving its inspection policies, but it also cautioned that the NRC needed to work harder to uncover "early indications of declining safety performance."[46] Advocates of nuclear power point to its fifty-year history of operation in the United States without any accidents significant enough to cause loss of life. Compare that, nuclear power advocates often argue, with more than forty thousand deaths in auto accidents in 2007.[47]

So how safe is safe enough? The debate goes on.

What About Terrorists?

Ever since September 11, Americans have had a whole new set of fears to contend with, and both the Nuclear Regulatory Commission and the Department of Homeland Security say they have increased their efforts to ensure safety at nuclear power plants. As one step, they asked the National Academies to determine how serious the risks were. The experts concluded that the chances that terrorists could somehow obtain nuclear material to make a "dirty bomb"* were "small, given existing security measures."[48] In fact, some would argue that nuclear power plants are more heavily monitored and better protected than other potential sources. The NRC notes that "radioactive materials are routinely used at hospitals, research facilities, industrial and construction sites," and that it has procedures for monitoring problems at these locations too.[49] Most terror experts believe the chance of terrorists obtaining nuclear material is greater outside the United States; and that the real danger is probably from a dirty bomb smuggled into the country.[50]

On the question of whether terrorists could attack a plant and damage it so that the radioactivity is released, the panel unanimously concluded that terrorists might be able to cause "potentially severe consequences," and recommended several changes that would ramp up safety, in-

---

* The uranium used in power plants isn't what experts call "weapons-grade," so it would have to go through difficult and expensive reprocessing before it could be made into a Hiroshima-style atomic bomb. A "dirty bomb" would pack a conventional bomb with uranium or other radioactive material. The blast wouldn't level a city, but it would spread the radioactivity over a wider area. Many experts, including the Federation of American Scientists, say a dirty bomb actually wouldn't kill very many people, even in a city. The real danger would be the subsequent panic.

cluding installing "water-spray systems" (which, curiously, sound like ultrasophisticated versions of hotel sprinklers).[51] In 2006, the GAO concluded that "nuclear power plants [have] made substantial security improvements in response to the September 11, 2001 attacks," but also noted that the NRC had not yet completed some planned inspections and pointed to several issues that "warrant NRC's continued attention."[52] For its part, the NRC points out that "power plants are among the most hardened commercial structures in the country and are designed to withstand extreme events, such as hurricanes, tornadoes, and earthquakes," and that they have "redundant safety systems" and "multiple barriers."[53]

Opponents of nuclear power question whether these measures are sufficient; worse yet, they believe, the NRC is radically underestimating the resolve and treachery of terrorists like those who planned the attacks of September 11. Dr. Edwin Lyman, representing the Union of Concerned Scientists, testified before Congress in 2005. "What I find most troubling," he said, "is that I see little evidence of 'outside-the-box' thinking going on in the NRC or in the industry in response to emerging threats or safety concerns. They do not want to question the assumptions they have made because they are afraid of the answers they might get, especially if those answers end up costing the industry more money. But I doubt that America's adversaries put similar constraints on themselves when plotting attacks."[54] The Natural Resources Defense Council is another group that believes that risks such as these, combined with the costs of nuclear energy, should effectively rule it out "as a leading means to combat global warming pollution."[55]

A FEW REALITIES TO PONDER

In many respects, the question facing Americans now is not whether to use nuclear power—we already do. The real question is whether we expand it, and how much we're willing to do and pay to minimize the risks that it poses, however serious we think they are. Many Americans see nuclear power as inherently unsafe and impractical, and, they point out, it's not as if the country doesn't have many other options. In their view, the nation's time, effort, and money would be far better spent investing in alternative and renewable energy sources and working much harder on energy efficiency and conservation.

But another reality is this: Nuclear power is one of the few low-carbon energy sources we have that operates all the time, rain or shine, dark or light. Whether we think about it much or not, we have lived side by side with nuclear power fairly uneventfully for about five decades now. France uses nuclear power extensively, also seemingly without much brouhaha. Yet, according to nearly every expert who has looked at the issue carefully, nuclear power has a very limited future in the United States unless we make it cost-competitive by taxing or fining carbon-dioxide-emitting fuels and finding ways to mitigate the massive investment needed to get a nuclear plant up and running. Is the country ready to do that to benefit a form of power so many people have doubts about?

Beyond this, we leave the issue for you to ponder.

Chernobyl and Three Mile Island: Twice-Told Tales

In the end, both accidents were less deadly than originally feared, but their names, Chernobyl and Three Mile Island, still shape public thinking about the safety of nuclear power. In the United States, the malfunction at Three Mile Island in Pennsylvania in 1979 prompted reform both in government oversight of nuclear power and in the industry's own practices and standards. The far more catastrophic disaster at Chernobyl in Ukraine in 1986 gave the world a lesson in the perils of nuclear power when placed in irresponsible, incompetent hands.

The two accidents are different in their origin and results, and it's worth considering what led to each and what they mean for humankind's ability to manage technology that can be extremely dangerous when mishandled.

Of the two, Chernobyl is by far the most sobering. In the words of the normally reassuring U.S. Nuclear Regulatory Commission, "the accident, caused by a sudden surge of power, destroyed the reactor and released massive amounts of radioactive material into the environment."[56] The explosion was powerful enough to blow a 1,000-metric-ton cover off the reactor.[57] The fires that followed could be seen from space.[58] Fifty tons of radioactive dust were scattered over 140,000 square miles of what is now Belarus, Russia, and Ukraine (then the Soviet Union), exposing nearly five million people to radiation.[59] Some 134 plant workers and emergency personnel suffered acute radiation poisoning or burns after the accident, and more than two dozen died.[60] Investigations after the accident suggest that Chernobyl will eventually lead to about four thousand cases of cancer.[61] One major consequence was an increased incidence of thyroid cancer stemming from young children and adolescents drinking contaminated milk following the accident, and the secrecy and incompetence of the officials made that problem worse. According to the World Health Organization, "if people had stopped giving locally supplied contaminated milk to children for a few months following the accident, it is likely that most of the increase in radiation-induced thyroid cancer would not have resulted."[62]

Apart from the eternal mystery of why human beings sometimes do dangerous and reckless things, there's no great mystery about what caused the accident at Chernobyl and why it was so deadly: human error. Plant managers and workers were operating the reactor in violation of the technical standards set for it and without proper safety precautions in order to test its performance at very low power.[63] They found out, all right—it became erratic and unstable and essentially blew up. Unlike nuclear reactors in the United States, the Soviet-designed reactor at Chernobyl had no concrete-and-steel containment shell to enclose or control the radioactive material after an accident.[64] Radiation flooded the area. Since plant managers initially tried to hide the danger, local residents were not warned to avoid contaminated food and milk. Evacuations didn't start until thirty-six hours after the initial explosion.[65] Pripyat, the town nearest the plant, is still off-limits to anyone without special permission.[66]

About a month after the accident, the radioactive debris was covered with a "sarcophagus" to shield the area from more exposure. Unfortunately, experts say that a sturdier, more durable, more impervious one is needed to protect against "the threat of the very large inventory of radioactive material contained within the existing sarcophagus."[67] A new structure designed to last a hundred years is now being built.[68] The accident at Three Mile Island was less deadly but still unnerving. The chain of events started at 4:00 a.m. on March 28, 1979. There was a problem in the "secondary, nonnuclear section of the plant," which led to a shutdown of the reactor. But the situation became much worse when "the instruments available to the reactor operators provided confusing information."[69] Although their gauges and meters told them that the reactor core was "properly covered with coolant," the vital fluid was actually draining rapidly through a stuck valve. The reactor overheated, and "the plant suffered a severe core meltdown, the most dangerous kind of nuclear power accident."[70]

As alarming as the situation was, and despite several tense days during which experts and officials scrambled to understand exactly what went wrong and how to handle it, some things at Three Mile Island actually went right.

- There was no "breach of containment," that is, the reactor's thick steel-reinforced concrete containment walls did not give way. What did happen is that some 700,000 gallons of radioactive cooling water flooded the buildings at the site, and some radioactivity escaped through a ventilating system.[71]
- Local and federal authorities went into emergency response mode quickly even though it was not immediately clear how serious the situation was. Federal response teams were dispatched to the scene within hours. Helicopters were in the air over Three Mile Island sampling for radioactivity by noon on the day of the accident.[72]

- Only modest amounts of radiation were released into the surrounding area. According to estimates from a whole series of governmental and nongovernmental studies, the average dose of radiation was about 1 millirem. As the NRC points out, you would get about 6 millirems of radiation from a series of chest X-rays.[73]
- No one died in the accident, and no deaths from cancer have been attributed to it. In 1996, a federal judge dismissed suits against Three Mile Island's owner, the General Public Utilities Corporation, by more than two thousand local residents who believed that their leukemia or other serious diseases stemmed from the radioactivity released in 1979. The judge wrote that there was a "paucity of proof" and "evidence insufficient" to show that the residents' illnesses were caused by the accident.[74]

Some Pennsylvanians and others remain convinced that Three Mile Island did lead to some cancers,[75] and some charge that the government has engaged in a cover-up.[76] Few disputes like this are ever resolved to 100 percent agreement, but the bulk of the official body of investigation suggests that the accident led to no major physical health effects. As the Nuclear Regulatory Commission puts it, "in spite of serious damage to the reactor, most of the radiation was contained," and "the actual release had negligible effects on the physical health of individuals or the environment."[77]

Three Mile Island did lead to some major changes in plant monitoring systems and employee training, since confusion among the staff and misreading of the reactor instruments unquestionably worsened the initial malfunction. The NRC lists more than a dozen important changes and reforms instituted to avoid a similar accident.[78]

But Three Mile Island stopped the U.S. nuclear industry in its tracks. For thirty years after the accident, no new plans for reactors were submitted. As the *Washington Post*'s business reporter Martha Hamilton put it, "it drove a stake through the heart" of the industry's future.[79]

For Americans considering whether nuclear power should play a stronger role in the nation's energy future, the fact that human error was a key factor in both Chernobyl and Three Mile Island is worth considering. The Chernobyl reactor was badly designed, true, but the human beings in charge took terrible risks that led to terrible consequences. For some it may be comforting to point out that this took place in

Ever want to see the inside of a nuclear reactor's containment shell? This is what one looks like. It helps prevent radioactivity from seeping into the atmosphere if there is an accident. The reactor at Chernobyl didn't have one. Photo courtesy of the Pacific Northwest National Laboratory

the former Soviet Union, never known for its efficiency. But wretched decision making knows no national boundaries. Just consider the terrible risks and wretched decision making America's bankers and financial geniuses have displayed in the last decade.

In the end, the dilemma remains: sometimes mistakes lead human beings to make improvements, and sometimes, when memories fade, people get careless and overconfident, and they make mistakes again.

This all comes down to the level of risk you're willing to take. The death toll at Chernobyl pales compared to the annual toll of traffic accidents, but nobody suggests we stop driving because of it. Of course, a traffic accident doesn't render an intersection poisonous to human life for decades, either. Other forms of energy have accidents too. Natural gas pipelines explode. Dams break (very rarely). Oil tankers spill. Nothing is 100 percent safe, ever. And human beings never do everything in every instance exactly the way they're supposed to either.

CHAPTER 10

AS LONG AS THE WIND BLOWS AND THE SUN SHINES

I'd put my money on the sun and solar energy. What a source of power! I hope we don't have to wait until oil and coal run out before we tackle that.
—*Thomas Edison, 1931*

Wind and solar power are as hip as technology gets in the energy world. Every clean energy commercial seems to include a row of solar panels glinting in the sun with Katrina and the Waves singing about "walking on sunshine" in the background or a wind turbine ticking over to the sounds of Bob Dylan singing "The answer, my friends, is blowing in the winnnnnnd . . ."

Wind and solar warrant this hype, at one level, because they are the fastest-growing sectors of the energy business. There was a 49 percent increase in photovoltaic solar installations between 2005 and 2006, and wind power is growing faster than any other energy source in America. The We Can Solve It initiative, backing Al Gore's call for "100 percent clean electricity in 10 years," is supporting these technologies big-time, calling on members to stuff flyers in with their utility bills and take other steps to pressure companies to convert to these systems. T. Boone Pickens, the oil billionaire, has invested in wind farms in Texas and spent major advertising dollars touting the idea. The Obama administration says incentives in its economic stimulus package will double the supply of renewable energy over the next four years.[1]

THE 7 PERCENT SOLUTION

Yet all of this hubbub is a little misleading. In fact, all "renewable" energy accounts for only 7 percent of U.S. energy use, according to federal statistics. And most of that comes from sources such as hydroelectric dams, ethanol, and biomass (wood-burning stoves and so on). Wind power is only 5 percent of that 7 percent, and solar is even less (1 percent of the 7 percent).[2] Or, in other terms, wind power accounted for 0.3 percent of all U.S. energy consumption in 2007, and solar only 0.1 percent. We're talking about fractions of fractions here, tiny slivers of total production. And the government's official energy projections expect renewables to be only 12.5 percent of our energy supply by 2030.[3]

At the same time, another branch of the Department of Energy projects wind power alone *could* provide 20 percent of our electricity by 2030.[4] Other countries have done much more. Wind power already provides 20 percent of the electricity in Denmark, 12 percent in Spain, and 7 percent in Germany.[5] By contrast, all the installed wind turbines in the United States together could supply about 1.2 percent of American electricity demand.[6]

Nowhere else in the energy field is the difference between *could* and *will* so great. Sure, it's possible to get a lot more energy from wind and solar—but it'll take work, money, and expectations that make sense.

THE PERFECT AND THE GOOD

Sometimes the debate on renewable energy makes it sound like all or nothing: Either we switch over to wind and solar or we pollute the Earth until it's unrecognizable. Critics tend to see wind and solar as "crunchy granola" options that are so limited that they are meaningless in terms of solving the country's energy problems. The sun doesn't always shine, and the wind doesn't always blow, so what do you do then? Most of us do expect to be able to turn on the lights on a windless night, after all.

There's an old line (apparently from Voltaire, but frequently recycled) that you shouldn't let the perfect be the enemy of the good. In other words, just because something isn't a complete and flawless solution doesn't mean it's not worth doing. That's certainly true with wind and solar energy. They're not perfect. They're not going to work every place, every time—at least not given the technology we're using now. We're not going to get all our electricity from them, and depending on where you live, you may never get much out of them at all. As long as we accept their limitations, however, that's not a problem. In certain parts of the country they can do a lot. There's no reason to dismiss these technologies because they can't do everything.

Half the battle on this is not overselling ourselves. For example, lots of people claim wind and solar will help the United States become independent of foreign oil, and that's not exactly true—at least not in the near future. Wind and solar are used to create electricity, and right now cars run on gas. Until we start using electric cars, wind and solar power aren't going to do anything to change the way our cars run, and so we'll still be importing oil or using alternatives such as liquefied natural gas or ethanol.

But both wind and solar have two huge advantages that

ought to be put right up front. First, they don't produce any greenhouse gases—in fact, they don't actually use "fuel" in any traditional sense at all.* Maybe they won't reduce our use of imported oil, but they could reduce our use of the fossil fuels that are used for electricity, namely, coal and natural gas. And that leads to advantage two, which is we're not likely to run out of wind or sunshine anytime soon. Let's take a closer look at each of the potential energy superstars.

INHERIT THE WIND

Wind power is one of the oldest energy technologies on the planet. Everyone knows about the picturesque Dutch windmills, and windmills have been a standard part of the midwestern farm scene for decades. That said, the modern wind turbine doesn't have much in common with those classic, picture-postcard windmills. Wind turbines now are huge, standing 250 feet high with airplane-style blades 130 feet long. (Taller turbines are better, because winds are stronger at about 250 feet above the ground. Longer blades are better because they generate more power as they spin.)

As we said, wind power is extremely clean: no fumes, no waste, no greenhouse gases. Once you put a turbine up, it just ticks over with some routine maintenance. No muss, no fuss. There are some environmental downsides, however. (Do you notice a theme here?) Wind farms aren't popular neighbors. Beauty may be in the eye of the beholder, but some people consider turbines on an aesthetic par with high-tension power lines. Additionally, turbines hum as

---

* Like virtually everything human beings do, there are some greenhouse gases produced along the way in manufacturing wind turbines and solar panels. But like nuclear power, the actual operation of wind and solar energy doesn't produce any greenhouse emissions.

they run, and the blades can cause "flicker effect" shadows as they spin that not only are annoying but also have been accused of interfering with airport radar. Finally, there's evidence that they're bad for migratory birds and bats, which tend to fly into them. The threat to birds and bats can be reduced, depending on how the turbines are designed and where they're located (it's best to keep them off ridges, for example).[7]

For wind advocates, these downsides don't sound like much compared to the risks of global warming, the troubled politics of oil, or the safety fears surrounding nuclear power. However, when the National Research Council looked at the issues surrounding wind power, they pointed out that there's no agreed-on standard for balancing the environmental trade-offs here—that is, for determining whether the local environmental problems outweigh the benefits of reducing greenhouse emissions.

We've seen this play out in one of the most ferocious local environmental fights in the country: that between Cape Wind, a company that wants to build a huge wind farm off Cape Cod, and local opponents who say the turbines will ruin Nantucket Sound. Obviously, the local opponents of projects such as Cape Wind think that the environmental problems trump the possible benefits (although many would argue the project would be okay if it was pushed 10 or 20 miles farther offshore).[8]

Cape Wind helps to illustrate three big issues holding wind power back:

Location, location, location. Just because there's wind where you live doesn't mean you can actually put up an effective windmill. In other words, while the wind blows everywhere, not everywhere is suited to a wind farm. It's not a matter of saying, "Hey, it's windy today. Let's put up a turbine."

The challenge in locating a wind farm is finding the sweet spot where the wind blows steadily and with power. An "excellent" wind, from a power generation point of view, averages about 17 miles an hour or more at 150 feet above the ground.[9] Plus, there's the issue of "density" (really—air has density). A dense wind has more "push" than a light one, and thus can spin a turbine more effectively. Generally speaking, cold air is denser than warm air, and lower altitudes are denser than high ones. The map on page 168 from the National Renewable Energy Laboratory identifying the best wind areas tells the story.

Overall, it's pretty logical once you think about it. If you've been on the beach or in a small boat, you know the wind blows more strongly offshore. So it's no surprise that there are big red bands of "superb" wind areas off the coasts and on the Great Lakes. That's why floating wind farms are a hot ticket right now, particularly in Europe, where land is at a premium.

As far as land-based wind farms go, the Great Plains, long the breadbasket of the world, is also the area best suited for wind power. It shouldn't astonish you, then, that Texas is the national leader in installed wind turbines.[10] If you look at the map, however, you'll see that entire regions of the country are pretty much out of luck when it comes to wind power. The Southeast has got nothing going on in wind power and probably never will. In fact, one of the biggest problems with wind power is that the wind frequently isn't where the people are. In fact, given both the wind patterns and the "not in my backyard" attitudes, the ideal location for a wind farm is usually known as "out in the middle of nowhere."

This isn't the end of the world. As we noted in Chapter 5, how you get your electricity is already heavily influenced by regional concerns. If you happen to live near a river suit-

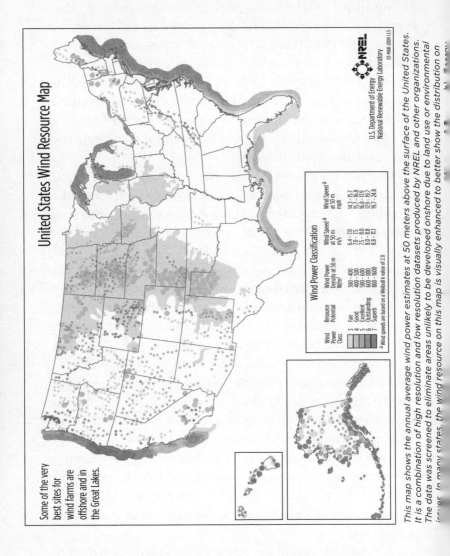

United States Wind Resource Map

Some of the very best sites for wind farms are offshore and in the Great Lakes.

Wind Power Classification

Wind Power Class	Resource Potential	Wind Power Density at 50 m W/m²	Wind Speed [a] at 50 m m/s	Wind Speed [a] at 50 m mph
3	Fair	300 - 400	6.4 - 7.0	14.3 - 15.7
4	Good	400 - 500	7.0 - 7.5	15.7 - 16.8
5	Excellent	500 - 600	7.5 - 8.0	16.8 - 17.9
6	Outstanding	600 - 800	8.0 - 8.8	17.9 - 19.7
7	Superb	800 - 1600	8.8 - 11.1	19.7 - 24.8

[a] Wind speeds are based on a Weibull k value of 2.0

U.S. Department of Energy
National Renewable Energy Laboratory

◆NREL

05-MAR-2009 1.1.5

This map shows the annual average wind power estimates at 50 meters above the surface of the United States. It is a combination of high resolution and low resolution datasets produced by NREL and other organizations. The data was screened to eliminate areas unlikely to be developed onshore due to land use or environmental issues. In many states, the wind resource on this map is visually enhanced to better show the distribution on

able for a dam—say, Nevada, the Pacific Northwest, or New England—you get clean hydroelectric power. If you don't, you don't.

But it's also true that utilities like to site new plants relatively close to where the demand is. They'd rather not have to build long-distance transmission lines to get wind-generated electricity from where the wind is blowing to where it could be of most use. Right now the states that use wind power most—Texas, California, Iowa, and Minnesota—are the ones that have good wind supply. It's fine if it stays that way, but if we want to meet that "20 percent by 2030 goal" set by the federal government, we'll have to spend $20 billion to build 12,000 miles of new transmission lines.[11]

Peaks and valleys. Someone once memorably sneered at the hard-drinking, hard-womanizing movie star Errol Flynn by saying, "You can count on Flynn. He'll always let you down." Critics of wind power say much the same thing. The wind doesn't always blow, and even when it does, it doesn't blow steadily. That's not the wind's fault, but it does mean that dramatically increasing wind power would be tough for our electricity grid to cope with, for two reasons.

While the existing power grid is designed to cope with swings in both supply and demand (demand surges on hot summer days, for example), those are relatively small and predictable. And fundamentally, they're based on what *consumers* want. Wind farms have big swings in output, and it's got nothing to do with what you want; it's about the weather.

With wind power, you can see swings of maybe 10 or 15 percent, and these come without much warning. It's possible to build a grid that can handle this without melting down; the Europeans are learning to cope with this. But the U.S. grid is getting old, and we're going to have to spend billions to bring it up to speed. If you saw any headlines about some government money being invested in the energy grid as part of President Obama's 2009 economic recovery act, it's true, but it's a very small step—not even 1 percent of the estimated costs.[12] (Read more on the power grid and why the United States needs a new one in "Grid and Bear It," page 188.)

Another way of looking at this is what engineers call "capacity factor," namely, the difference between how much energy a power plant usually produces and how much it could be producing running full tilt. No matter what the power source is, there's a gap between theoretical performance and actual performance. Nuclear power plants are almost always running flat out, with a capacity factor of 90 percent. Coal plants have a capacity factor of 70 percent. Wind turbines, however, only have a capacity factor of between 20 percent and 40 percent, depending on where they're located.[13]

Which leads us naturally to the freakiest drawback of all—that we may need just as many conventional power plants as we ever did. How can this possibly be? you ask, after seeing all those commercials for wind power. Because there simply has to be standby, backup power for consumers to use when there's no wind blowing.

In a way, it helps to think of wind power as much like a hybrid car. In a hybrid, there's an electric motor that powers the car at low speeds, and then when you reach a certain speed the conventional gasoline engine kicks in. You still need a gasoline engine powerful enough to move the car at highway speeds; the advantage of a hybrid is that you don't use that gas engine as much. If you mostly drive short errands around town, you won't use the gas engine much at all, and you save fuel and release smaller amounts of greenhouse gases into the atmosphere.

Wind power is the hybrid of electricity generation. On days with enough wind, the turbines will spin and produce electricity. In fact, on days with strong wind they may actually produce more than you need. On calm days, you need a conventional plant (and right now that would be one powered by coal, nuclear, or natural gas) to pick up the slack.* But just as your hybrid car needs a gasoline engine powerful enough to move the entire car, your electric utility needs a conventional power plant powerful enough to run the entire town. However, with wind power, you just won't run the conventional plant as much.

Potentially, this saves fuel, and cuts greenhouse gases. But it does mean that utilities have to build conventional plants to cover the full demand for electricity, just as if the wind farm didn't exist at all. This doesn't always get factored into the capital costs of wind farms, but it should. The more we turn to wind power, the more important this baseload capacity is. In Texas, the state electricity council has already had to put its emergency plan into service at least once, cutting power to some customers because wind generation suddenly dropped.[14] In 2007, a Montana utility built a new natural-gas plant just to compensate for wind farms. (Most experts say natural gas is a good option for these baseload capacity plants, because it fires up quickly. Nuclear and geothermal plants have the same advantage).

This is one of the main reasons why wind power is actually a pretty expensive option for producing electricity. Even though you don't have to pay for the "fuel," the initial capital costs are higher than coal or natural gas when they're spread out over the life of the plant (or "levelized," in

---

* Engineers are working on improved batteries that can save the excess wind energy for later, but right now we can't store this electricity on the vast scale needed.

energy-speak).[15] A study by the British Royal Academy of Engineering found that when the cost of standby generation is included, wind farms were actually among the most expensive options for electricity.[16]

HERE COMES THE SUN

Not to get all *Beakman's World* about this, but from a scientific point of view, all energy comes from the sun. No sun means no warmth, no plants, no life. Period. All of our other forms of energy are essentially repackaging the solar power that allowed them to exist in the first place.

That said, using solar power specifically is also an old technology. The Romans used solar collectors to heat their baths, and solar hot water tanks were a common fixture on California rooftops as recently as a hundred years ago. In 1954, Bell Labs invented the photovoltaic (PV) cell, which converts sunlight directly into electricity. NASA loves solar cells. Those big winglike things you've seen sticking out of satellites and space vehicles are solar panels, and they've been the space agency's preferred power system for a half century.

One of the great virtues of solar is that it can operate in very rugged conditions. When international aid agencies try to come up with solutions for extremely poor sections of the Third World, more often than not they turn to solar. (Remember, more than 1.6 billion people on this planet don't have electricity. Whatever power they get comes from biomass: wood, peat, and other stuff you can throw on a fire.) Solar cells power satellite phones and even televisions in the developing world, and simple metal solar ovens are widely used in Africa.

Another virtue of solar is that it can be an individual power source. If you're interested in living greener and can afford it, you can hire a contractor and put your own solar panels up on the roof to power your house. Obviously, that's not something you can do if you prefer nuclear power. There were some thirty-nine thousand residential photovoltaic installations linked to the power grid in 2007, and another nine thousand commercial and industrial sites. Since the business sites tend to be bigger, they account for two-thirds of the actual solar electricity generated, even though there are fewer installations.[17]

There are three basic kinds of solar power, and the options for generating it pretty much go back to either the Romans or NASA:

The NASA Version: Photovoltaics

As we said, these create electricity directly when sunlight interacts with silicon semiconductors in the panel. An individual cell can produce anywhere from 10 to 300 watts, but you can also hook a number of cells together into arrays of almost any size. The most common use is to provide power to an individual house or building, but you can also build a power plant with these.

Even prisons can go green. A solar thermal system provides hot water at the Jefferson County Jail in Golden, Colorado. Photo by Dave Parsons. Courtesy of the Department of Energy, National Renewable Energy Laboratory

The Roman Version: Solar Hot Water

You can still buy these, and while they used to be huge, ugly tanks, now they're comparatively thin water conduits and roughly about the same size as photovoltaic panels. Unless you're a fan of cold showers, these probably won't completely eliminate the need for a hot water heater in your house, but they'll cut down on how often you use it. (Remember the hybrid car analogy?) There are also installations out there specifically to heat swimming pools.

Getting Up to Steam: Solar Thermal Electric

This is the preferred option for the few large solar power plants out there. Solar power is used to create hot water and steam, which are in turn used to run turbines and produce electricity. Eleven such plants are operating in the Southwest, all but two built during the 1980s and early 1990s, the last time when fuel prices were high.

WHERE THE SUN DOESN'T SHINE

You're far enough along in this book now to know that every energy source has a "but." Solar's no exception. In fact, like wind, it's got at least three big drawbacks.

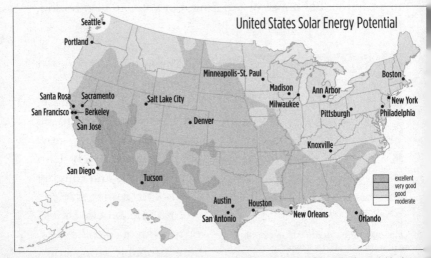

The sunny southwest offers the best solar potential. To no one's suprise, Seattle and Alaska are less promising. Source: U.S. Department of Energy, Energy Efficiency & Renewable Energy

Your Toaster Doesn't Work at Night

This is pretty obvious: no sun, no solar power. Still, even if you live in famously rainy Seattle, it's not like the sun *never* shines, and most of North America is actually fairly well suited for using solar power. The hot spots for solar, if you'll forgive the term, are in the Southwest, such as southern California, Arizona, Nevada, and New Mexico. No surprise there. But if you take a look at the map on page 174, you'll notice that it's quite a bit more optimistic than the wind power map. Most of the country can make some use of solar power, even places such as Seattle, Portland, and Alaska. (Skeptics who have visited the Pacific Northwest repeatedly and never seen the sun shine will just have to accept the experts' word on this.)

The sun is at its best and brightest, at least from a photovoltaic point of view, in the middle of the day, which makes perfect sense. And if you're powering an office building with solar, that's no problem. Daylight hours and business hours line up just fine. But since reading by candlelight has fallen out of favor, most people expect the electricity to be up and running when they come home from work and switch on the TV. Simply put, when the sun isn't shining, you're out of luck. You're not getting anything at night; you're not getting anything (or not much) when it's raining. Plus, in its own ornery fashion, the sun may be shining exactly when you don't need it.

As drawbacks go, this isn't that challenging. Some home owners who use solar get around this by using storage batteries, but even more deal with this by linking up to the existing power grid in their community. Essentially, if you put solar panels on your roof and you're part of the grid, you become a mini power plant yourself. At times of day when you're generating more energy than you need (say, around noon), you sell that power to the local electric utility, which sends it to somebody else. At times when you're not generating enough power (like after dark), the grid sells power to you to run your house.

Sounds pretty simple, and it is, at least in the thirty-five states where utilities have adjusted their rules to allow for this. The term to look for is "net metering."[18] When net metering is in place, not only is the local utility willing to buy your surplus electricity, it also agrees to pay you the retail price—in effect, the utility pays you the same rate it charges you for the power it provides. When you're pumping out excess electricity at lunchtime, your meter "runs backward," as the experts say. That's significant, because conventional power plants charge the utility a *wholesale* price, which is less than what's on your electric bill. Without net metering, you'd lose money on the deal.

Where Are All These Panels Going to Go?

Critics of solar often raise the issue of how much land it would take to host significant amounts of solar panels. Generally speaking, it takes 5 to 10 acres of PV panels to generate 1 megawatt of electricity.[19] To produce all the electricity the United States needs using photovoltaic cells would take 10 million acres of land, according to the federal Energy Department.[20]

Whether you think that's a lot or a little seems to depend on your point of view. Solar advocates say that's less than half the acreage in the Arctic National Wildlife Refuge—but given the controversy over that project, it's not clear whether that argument makes things better or worse. And for what it's worth, roofs count, so you're actually getting double duty out of some land. In Spain, one town installed solar panels in the local graveyard. Apparently, neither the living nor the dead complained there, but we wouldn't count on that here.[21]

A number of the most promising areas for large solar

plants are on federal lands in the Southwest, and as of 2008, the Bureau of Land Management had more than two hundred applications to build plants. California alone had eighty applications that would cover more than 700,000 acres.[22] The BLM is trying to come up with a policy on how to review the environmental impact of these projects, and until it does, this is going to move slowly.[23]

Solar Costs More than Other Energy Sources

Everyone agrees this is the killer issue here. Even when you average it out over time, electricity from photovoltaic cells costs anywhere from two to four times as much as the national average for conventional electricity.[24] That's on an average basis, however. Given that people in different parts of the country pay widely different rates, solar is already competitive in some areas.[25]

The problem is mostly in the installation costs. Just like wind power, once you have a solar system in place, your only major cost is maintenance—you don't have to pay for fuel. But while the technology is getting better and cheaper, the installation costs are still steep. The average size for a residential installation in the United States is a little under 5 kilowatts and can range from $30,000 to $40,000 to install. (It hasn't helped that there's been a worldwide shortage of silicon over the last few years, thanks to increased demand for everything from solar systems to computer chips.)[26]

Because solar isn't cheap, it's most popular in places that have high conventional electricity bills—and strong state government incentives in place to make solar cheaper. California, for example, has five times the national average of photovoltaic installations per capita. Up until 2006, when federal incentives changed, Hawaii had half the solar hot water heaters in the nation. In both cases, these were states with extremely high energy costs (Hawaiians paid more than 24 cents per kilowatt-hour for electricity in 2007, six cents more than the next-highest state, Connecticut.) Effectively, solar is competitive in these states because regular electricity is pricey *and* the state has incentives to make solar cheaper.[27]

WANT TO BUY THIRTY YEARS OF ELECTRICITY UP FRONT?

If every other drawback and objection could be overcome, wind and solar would still face one simple issue: they're just more expensive than other alternatives. And while the specific technical issues are different, the pattern is the same, in that high initial costs cancel out the fact that the "fuel" is free. Fully half the cost of a wind turbine, for example, is the actual construction. By contrast, a coal or natural gas power plant is comparatively cheap up front, and even when the fuel costs over time are factored in, coal plants still end up being cheaper. "Solar is the ultimate capital good, and historically the way we've been selling solar panels is to ask the customer to basically pay their utility bill for the next thirty years all at once," says Charles Gay, corporate vice president and general manager of Applied Materials' newly formed Solar Business Group.[28]

So why does anyone buy renewable energy at all? "Utility market research has found that for residential customers, the decision is predominantly an emotional one, while for commercial customers, it is a business decision," according to the National Renewable Energy Laboratory. In other words, people buy wind turbines and solar cells because they want to do the right thing.[29] That's admirable, but nobody thinks nobility alone is going to bring renewable energy to the next level. If it's going to take off, it's got to be economically competitive with other forms of power.

In fact, that was the goal of the Solar America Initiative, unveiled by the Bush administration in 2006: to make solar power competitive with grid prices by 2015. Wind advocates have similar goals.

There are a couple ways of making that happen. One is that the technology itself gets cheaper, and on this front there's been progress. A photovoltaic cell costs half as much to build as it did in 1980, for example.[30] As with many products, the more PV cells you make, the more cost-effective it gets. At least one study argues that doubling photovoltaic production would bring the per-cell cost down by 20 percent.[31]

Another way—and most renewable energy advocates are really counting on this, whether they admit it or not—is that conventional electricity prices keep rising. This is why many of them support a carbon emissions tax or a trading system where companies have to purchase carbon permits (more about these ideas in Chapter 15). These plans would make electricity generated by coal and natural gas more expensive, so that wind and solar would have a fighting chance. Another approach would be to require coal-fired electricity plants to use expensive new technology to remove the carbon dioxide before it's released into the air. Same end result: electricity from coal would be more expensive and wind and solar would look better by comparison.

The third way of attacking this is for the government to offer more incentives—to make it worth people's while to go with wind or solar power with tax credits and other financial advantages. Unfortunately, the federal government has been anything but consistent on this. The United States does offer a renewable-energy tax credit, which Congress has let expire and then reinstated no fewer than three times in the past decade. The credit was extended again as part of the Wall Street bailout in 2008, but the on-again, off-again credit has contributed to a boom-and-bust cycle in the renewable energy business.

State governments seem to be less flighty. Some thirty-four states now have targets for renewable energy, calling for specific percentages of their state's electricity to come from clean sources.[32] California has been a model in this regard, and the California Solar Initiative is designed to offer generous incentives over the next ten years to make solar power competitive with other sources.

Without consistent government effort to permanently make fossil fuels more expensive (or alternatives cheaper), energy prices keep bumping up and down, and this puts wind and solar on a permanent roller-coaster ride. As oil prices spiked in 2007 and 2008, everybody wanted to get in on renewable energy. And when the global financial crisis hit and prices fell again, everybody wanted to get back out again. (Never mind that, as we pointed out in the beginning, wind and solar power don't have much to do with how we use oil.)[33]

That's the bottom line with wind and solar. If conventional energy prices get higher and stay higher, then these may take off as energy sources on their own. But if energy prices are relatively low—or even if they keep bouncing up and down—these will be competitive only if we as a society decide to *make* them competitive.

Dam It: Hydroelectric Power

Hoover Dam is probably the most famous dam in America. It's taller than the Washington Monument and contains enough concrete to build a 16-foot-wide highway from San Francisco to New York.[34] By any measure it's a vast, impressive construction. Yet the word *dam* itself always gets third graders giggling. We can think of no better way of solving this than by getting this quote from *National Lampoon's Vegas Vacation* out of the way: "I am your dam guide, Arnie. Please don't wander off the dam tour, and please take all the dam pictures you want. Now are there any dam questions?"

That said, Hoover Dam wasn't built to provide comedy; it was built specifically to produce electricity—4 billion kilowatts' worth, enough to supply 1.3 million people in three states.[35]

There are about 2,200 power dams in the United States, ranging from giant facilities to small plants designed to run a single factory. They are an attractive option because once you get past the pretty expensive initial cost, hydroelectric is cheap to run. From a climate-change point of view, hydroelectric power is outstanding, producing no greenhouse gases. It is also the nation's largest renewable energy source, providing 8 percent of all electricity in the United States—far more than wind and solar combined. On the Pacific Coast, hydro provides more than 40 percent of all electricity.[36]

Despite its attractions, hydroelectric power's share of total electricity generation is actually shrinking, from 30 percent in 1950 to less than 10 percent by 2000. Construction of major hydroelectric projects has basically stopped in the United States, and even optimistic projections don't see hydro's share increasing much in the next twenty years.[37]

There are only so many rivers in North America suitable for power dams, and the best of them are already done. The first hydroelectric plant opened in 1882, but the real era of dam building was in the first half of the twentieth century, which gave us huge projects such as Hoover Dam, the Grand Coulee Dam, and the massive Tennessee Valley Authority. Since then, even though electricity demand has gone up, hydroelectric production has largely stood still.

That's not to say we couldn't do more. A federal assessment says there are nearly 5,700 additional sites nationwide suitable for hydropower that could generate another 30,000 megawatts of power. The existing 2,200 hydroelectric plants produce 80,000 megawatts a year.[38] So those potential locations are smaller, and the power generated, while useful, won't make a huge dent in demand. In fact, it wouldn't even double hydropower's small share of the electricity market.[39] There's also research into making existing dams more efficient, with new turbines that produce more energy with less water.

But the major hurdle to hydroelectric power in the United States is often intense local and environmental opposition. While hydroelectric power doesn't produce any greenhouse gases, damming up a river has major consequences for fish and wildlife, changing wetlands and fish migration patterns. You can mitigate those effects by installing "fish ladders" to let salmon get around the dam, for example, but this doesn't always happen. American Rivers, an environmental group, points out that older dams often lack these environmental protections and utilities often resist them to keep down costs. According to the group, much of the country's hydroelectric power "has been run for generations without modern environmental protections. The impact has been devastating."[40]

Some argue that dams often aren't worth the environmental damage. Since many older U.S. dams need repairs and renewed federal licensing, these licensing reviews often require upgrades to ease environmental impacts that weren't imagined when the dam was built. In some cases, such as the Condit Dam on the White Salmon River in Washington state, the local power company decided it would be cheaper to spend $17 million to remove the dam than $30 million to $50 million to meet federal standards.[41]

The good news is that dams aren't the only form of hydroelectric power. If you're lucky enough to live right on a suitable river, you can actually buy a microhydroelectric system that can power a home or farm (these are considered very promising options in the developing world). Naturally, that's not going to make a big dent in the country's overall energy needs. Someday, however, harnessing wave and tidal power might be. These "hydrokinetic" generators utilize the ebb and flow of wave motions to make electricity. There are a number of pilot projects across the country, including the Roosevelt Island Tidal Energy project in New York's East River. These are experimental technologies, however, and it'll probably take decades before they produce significant amounts of power.

Hydro is going to keep generating power, and the country is going to have to make some decisions about investing and upgrading the dams we have. But until new technologies come online, we're not going to see a lot more from hydropower.

So there's a reason why you never hear anyone chanting "Dam, baby, dam!" "Damn, baby, damn" is more like it.

Looking for the Newest Hot Spots: Geothermal Power

In the world of renewable energy, geothermal power is often an also-ran, or maybe even a never-mentioned. Yet if you've ever been to Yellowstone or another geyser field, you know that this power source is awesome in the original sense of the word: "inspiring reverence, fear, and wonder." The first trapper and guide to come across the Geysers in northern California in 1849 compared them to the gates of hell. And since a glance at human history proves that damnation is a renewable resource, the idea of running a pipeline to hell seems likely to tap a lot of energy.

In reality, geothermal power harnesses the heat generated by the planet itself—not from hellfire, as that early trapper might have imagined, but from the superhot magma under the earth's crust that heats water underground. If the geology's right, you can drive a well down and tap into hot springs or even ready-made steam. Once they're tapped, you can use them either for heating or to run turbines to generate electricity. Like other renewables, you can harness this in your own home or office by getting a geothermal heat pump, which brings heat up from underground and can cut your heating bills by up to 75 percent. Some communities, such as Boise, Idaho, have used geothermal in district heating systems for a century. And the best part of geothermal energy? Since the earth's molten core isn't likely to cool off anytime soon, we're not likely to run out.

If we're talking about running electric power plants using geothermal power, the Italians led the way by tapping into the

Larderello steam field back in 1904. Similar plants are running at likely spots around the world in countries with natural geysers, such as New Zealand and Iceland, and at the Geysers north of San Francisco, which produce enough electricity to run five California counties.

Right now, geothermal is only a tiny sliver of U.S. energy production, accounting for about 2,500 megawatts of electricity annually. That sounds substantial, but one large nuclear plant by itself can generate 3,825 megawatts annually.[42] Put another way, geothermal provides a tiny 0.3 percent of all the energy in the United States.[43]

Lately, energy heavy hitters as diverse as Al Gore, Chevron, Google, and MIT have started pushing for more geothermal energy, and there's more geothermal power to be had in the United States, according to the U.S. Geological Survey. There are 241 known geothermal sites in thirteen states, mostly in the West, defined as places that could supply heat at 194 degrees Fahrenheit (90 degrees Celsius) or better. These places could provide another 9,000 megawatts a year if developed. The USGS also estimates that further exploration could identify sites that might provide another 30,000 megawatts.[44]

If you've been reading the rest of the chapter on alternative energy, you probably know where this is going. Like wind or hydro power, geothermal is a regional resource. With the geothermal technology available now, you have to drive your well at a "hot spot," which may be a long way from where the actual energy customers are. Also, like wind, solar, and hydro, the up-front costs are very high, even though the fuel costs are negligible. Geothermal is also great from a climate change perspective: The plants do release some steam, but emissions overall are minimal.

With a relatively small investment (say, about $800 million to $1 billion), MIT scientists concluded, geothermal could provide 100,000 megawatts, or 10 percent of the nation's baseload electricity by 2050. That may not sound like very much very soon, and other experts are more optimistic. But if we're going to make greater use of wind and solar power, having reliable, clean baseload power is critical. So geothermal could be a useful building block in getting the most out of other alternative sources.[45]

MIT argues that the real promise of geothermal depends on bringing online a new technology known as enhanced geothermal systems. What this boils down to, if you'll pardon the expression, is a system that pumps water down into the ground, heats it up using the earth's heat, then pumps it back up again to run turbines. Then, instead of drilling wells looking for a pool of hot water that's already there, you can make your own—which means you could use geothermal almost anywhere. There are successful experiments with this technology now in France and Australia.

The idea of enhanced geothermal does worry some people, however. A geothermal project in Switzerland was shut down after being accused of causing an earthquake in 2006.[46] Opponents are already citing the Swiss case to try and block a new geothermal project north of San Francisco. In addition, until recently geothermal has gotten a cold reception from policymakers. In 2006, the Bush administration proposed cutting geothermal funding entirely, arguing it's a "mature technology" that doesn't need government help. The Obama administration, however, is putting economic stimulus money behind geothermal.

Like other alternatives, the cost factor is critical for geo-

thermal. For it to be a significant energy source, the price of fossil fuels will have to be high—or the government is going to have to do something to encourage it to happen. The question is whether geothermal will have the backing to go from hot idea to hot property.

Grid and Bear It:
Why the United States Needs a
New Electricity Grid

As a discussion topic, the electricity grid combines all the worst qualities of the tax code and your local water and sewer authority. Go ahead, try starting a conversation about it with someone sometime. (We've tried.) Just like the tax code, it's deadly dull, yet touches every aspect of your life. And just like the water and sewer mains running under your house, you only think about the electric grid when it breaks.

By that standard—the "I'll think about it when it breaks" rule—you can expect to hear a good deal about the grid over the next few years. Most of the debate over energy is about how we generate it; very little has been about how we send it around. Yet this vital piece of infrastructure is getting old and technologically crotchety, and the demands on it are only going to grow over the next few decades. "It's becoming increasingly clear that we are operating the system closer to the edge than in the past," says Rick Sergel, the president of the North American Electric Reliability Corporation—in other words, the guy who's paid to worry about this stuff.[46]

Just as important, we can't do much to overhaul our energy policy without improving the transmission grid. Among the things the experts say *can't* happen unless the grid gets a makeover are:

- **Expanding wind and solar power.** If you looked at the rest of this chapter, you know that these energy sources (wind in particular) are usually in the middle of nowhere, far away from most customers. We need better long-distance lines to carry the power to where it's needed. Plus, the grid needs to be able to deal with the peaks and valleys in power generation that occur naturally when the wind dies off or the sun goes down.

- **Moving to electric cars.** If plug-in hybrids take off, that means switching over millions of cars from liquid fuels to drawing power off the grid. With luck, those cars will mostly be drawing power overnight, when demand is otherwise low. But at a minimum that means an increase in demand (8 percent by some estimates), and depending on how the change happens, it could mean making electric charging stations as common as gas stations are now.[47]

- **Switching from coal to natural gas or nuclear power.** Closing old plants and starting up new ones requires putting in new transmission lines (and pipelines if you're using natural gas), not to mention a system capable of balancing the load of new electricity sources.[48]

- **Just keeping up with demand.** Already there are congested areas in southern California and the New York–Washington corridor where the system is running close to capacity.[49] With demand projected to grow, we'll need more transmission capacity just to keep up, much less change to new forms of energy.

So what's wrong with the grid we have? Nothing, really, other than it was designed for a different era, and we haven't been willing to spend the money to keep up.

The basic structure of the grid was set up in the 1920s, an era where electric power was a purely local business, and a monopoly as well. Huge regional utility companies generated the power, sent it through their own transmission network to your house, and sent you a bill. Demand was pretty predictable and so was supply, so coping with plant outages or surges in demand was fairly straightforward. The service was pretty good, but if you didn't like it, tough (although these companies were and are heavily regulated by state governments). Those companies are still around, and still big players. The so-called vertically integrated utilities account for about 75 percent of the electric generation and customers in the United States.[50]

Over time, the utilities started sharing power with each other to meet demand and cover for each other during outages. That's still a regional phenomenon, though. Utilities in Oregon and Washington, for example, can share power, but they can't send or get any from the East Coast.

Even though it's still regional, the U.S. transmission network is huge by any standard, with more than 220,000 miles of high-voltage power lines (the big metal towers) and another 5 million miles of distribution lines (the wooden poles outside your house).[51] We're going to have to start replacing it anyway (30 percent or more of the grid is forty to fifty years old). It isn't just a question of buying new stuff, either. As we've mentioned before, energy demand is increasing worldwide, which means China, India, and everybody else is in the market for new grid technology too. Some experts are predicting a global backlog for electrical equipment, which will make upgrading more difficult.[52]

So what kind of grid do we need? Partly, we need to physically put in new wires and transformers. Right now an enormous amount of electricity evaporates in transmission—what the industry calls "line loss." Upgrading transformers and power lines could save 6 percent of all the electricity generated in the United States. Most experts are touting the idea of a two-way "smart grid," one that uses information technology to monitor use at both the utility's end and the customer's end as well.

Think of it as combining the communications power of the Internet with the brute force capacity of the transmission network. For example, right now the utility knows how much power you're using, but not necessarily why you're using it. A smart grid would allow the utility to know much more precisely what was causing a surge in demand, and shift resources to meet it. And technology at your business or home would know the grid was approaching capacity and be able to make its own adjustments, cutting back on the electricity you could afford to give up (say, adjusting your thermostat) while keeping the essential stuff. Another plus is that instead of that cryptic electric meter out back behind the rosebushes, you'd have a monitor in your home that would tell you in real time how your home or business is using electricity. Studies in Britain and Canada have found that simply giving people such information can reduce home electricity use by 5 to 15 percent.[53]

The problem is that none of this stuff is cheap. In fact, some estimates say we need to spend more on transmission than on new power plants. A study conducted for the Edison Electric Institute (the trade association of utility companies) estimate that the United States needs to spend $1.5 trillion to $2 trillion on upgrading its electricity system by 2030, which works out to $880 billion for transmission and distribution and $505 billion on new generating plants.[54]

Historically, the utility industry has paid the freight for this, but everyone agrees what they've been spending doesn't cover what's needed. In recent years, utilities have spent about $3.5 billion to $5 billion per year on transmission, and that's not going to be enough.[55] Utility companies complain that the fact that the United States hasn't had a clear energy policy makes it difficult for them to plan: Are they going to have to get off coal and switch to gas, or not? Is the government going to make fossil fuels more expensive, or not? Plus, new high-tension wires are no more popular than power plants themselves, and frequently face strong community opposition. The power industry would like the federal government to step in more and force communities to accept new lines, but that hasn't happened yet (at least as this book goes to print).[56]

The grid underlies everything the United States does, or wants to do, with energy policy. So despite its inherent boringness as a topic, unless we get the grid right, we're likely to get everything else wrong about energy.

CHAPTER 11

NO PLACE LIKE [AN ENERGY-EFFICIENT] HOME

Our house is a very, very, very fine house, with two cats in the yard.

— *"Our House," by Graham Nash*

Graham Nash's cats probably didn't use much energy—most cats don't like long, hot showers, and they rarely leave the lights on. The human beings in the house might be another matter entirely. Nash apparently wrote the song when he was living with Joni Mitchell, and both performers consider themselves environmentalists. We'd like to think they were reasonably careful about using energy, but even Americans who are conscientious still consume a lot. In fact, did you know that:

- About a fifth of the country's energy goes to residential use.[1]
- Residential and commercial buildings combined account for about 40 percent of consumption and a similar amount of greenhouse gas emissions.[2]
- Buildings actually edged out transportation in the emissions category—which is saying something.[3]

Obviously, we have some control over how much energy we use in our homes and offices, and (since you're reading a book on energy) you probably already know most of the dos and don'ts. Maybe you're even feeling a guilty impulse now to get up and check the thermostat. Our mission here isn't to remind you about running the dishwasher only when it's full. That's helpful, but it's not going to be enough. The reality is that to tackle our massive energy challenge, the United States needs to make some more far-reaching changes in the buildings we live in, what we put inside them, and maybe even where we build them.

In this chapter, we'll review some of the ideas planners, architects, builders, appliance manufacturers, and elected officials are proposing to save energy. Some of them are amazingly simple, if only we would do them. Some of them chip away at some cherished visions of how we live. Even so, we need to take a serious look at them if we're going to get a grip on the nation's energy crisis.

A COUNTRY THAT LIVES BIG

On modern TV sitcoms, lots of people seem to live in apartments—Will and Grace, the gang from *Friends*, and the *Sex in the City* foursome are just a few.* But when most Americans think home sweet home, they're not envisioning a tidy little one-bedroom, third-floor walk-up. They're thinking *Leave It to Beaver*: the single-family house with a dining room, family room, spare bedroom, and big yard on a nice leafy street lined with other single-family homes. The picket fence is optional.

In the funhouse mirror of TV land, even the dysfunc-

---

* Whether they could actually afford the apartments they live in is a question for another day.

tional families live more comfortably than rich people throughout most of history. The Bundys of *Married with Children*, the Simpsons, Raymond and his quarrelsome clan, and even the *Malcolm in the Middle* family (which didn't seem to have a last name) all live in spacious, if not always well-kept, two-story suburban houses with big kitchens and nice appliances. The message to people worldwide and right here at home? In the United States, you don't need to be accomplished or even especially well-off to have a big house with lots of stuff in it. It's the American lifestyle.

In reality, we don't all get to live in our Donna Reed dream house, but according to statistics from the Department of Energy, the average U.S. house is not too shabby. It was built in the late 1960s, and it has three occupants, more than 2,000 square feet of heated floor space, three bedrooms, two baths, central air-conditioning, three color televisions,* and three ceiling fans.[4] Impressive. It's why people worldwide want to live like we do.

The hitch is that our way of life eats up massive quantities of energy. With about 5 percent of the world's population, the United States is consuming nearly 22 percent of the world's energy.[5] This is not sustainable, and we are not going to be able to get a grip on the problem unless we develop a more manageable lifestyle. Part of that may mean reconsidering our expansive, living-life-large ideas about our homes.

Here are some questions to think over.

---

* The federal government still uses the quaintly specific term "color television" in its statistics, even though these days just about the only place you can find a black-and-white set is on eBay (www.census.gov/compendia/statab/tables/09s1090.pdf).

SHOULD WE BUILD OUR BUILDINGS DIFFERENTLY?

According to the Department of Energy and the Environmental Protection Agency, small changes in the way typical houses are built could make them 20 to 30 percent
more energy-efficient.[6] The agencies' joint Energy Star
Home program promotes these design and construction
changes nationwide. (Energy Star has a program for commercial buildings as well.) The Energy Star website (www
.energystar.gov) has a clever "Behind the Walls" feature
showing how better insulation, ducts and windows that
are more airtight, and caulking gone wild can keep the
heat in and the cold out, and thus save energy. The recommendations can also keep some creepy-crawly creatures
out too, another plus.* Going the Energy Star route not
only lowers energy consumption and greenhouse emissions
but actually saves home owners money in energy bills over
time.[7]

So far, Energy Star is voluntary, and about 940,000
new homes have been built to its standards since it started.[8]
But it's slow going. In 2008, about 17 percent of newly built
houses complied with the guidelines.[9] That leaves plenty of
room for improvement, and some communities have tried
to solve this problem by making compliance the law. New
houses in the Long Island (New York) communities of
Brookhaven, Riverhead, Babylon, Hempstead, Huntington,
Southampton, Oyster Bay, and others have to meet the Energy Star standards to get a certificate of occupancy.[10] Obviously, there are some additional up-front costs involved
in meeting the new building codes,[11] but three-quarters of

---

* We think the Energy Star people ought to push this benefit more. If
you've ever had an encounter with "a spider the size of a Buick," to
quote Woody Allen, this is a very motivating idea.

Americans say developers should be required to build more energy-efficient homes, even if they are more expensive.[12]

The downside? Well, there are those up-front costs. When an Energy Star law was debated in Shelter Island, New York, some architects raised concerns about the approval process, complaining that the inspectors who give the thumbs-up or -down on whether a building has met the criteria generally have less training than the architect.[13] Beyond that, some experts argue that many builders don't even comply with current codes. Michael DeWein, an expert from the Building Codes Assistance Project, believes the United States could probably get a 30 percent energy efficiency saving just from enforcing current laws.[14]

Some ideas about changing the way we build buildings are more ambitious. One is to create zero-energy buildings (or ZEBs, as the Department of Energy dubs them).* It's a lot like it sounds. A ZEB, for example, might have enough solar panels to produce all the energy the building needs; maybe it even sends some extra energy back to the community for someone else to use. Right now, ZEBs are pretty rare, mainly larger public buildings or at colleges and universities, such as Oberlin College's Lewis Center and the Science House at the Science Museum of Minnesota.[15]

IS IT TIME TO BECOME A RENOVATION NATION?

Changing the way new buildings are built may help, but most of us aren't moving into brand-spanking-new homes.[16] About half the building space in the country will need renovation over the next thirty years anyway, so is this our big chance?[17]

Again, some of the changes are relatively simple. Ac-

---

* Somewhere deep in the bowels of the Department of Energy, there must be a deputy assistant undersecretary for acronyms.

cording to the Energy Star people, sealing the cracks that let cold air in and heat out, adding insulation, and replacing older furnaces and air-conditioning units can reduce consumption and emissions and save owners and landlords money over time.[18]

There are tax incentives tempting us to make changes like this. Plus the Economic Recovery Act of 2009 includes about $8 billion for energy conservation and weatherization, both to tackle the country's energy problems and to provide jobs.[19] To get this kind of work done, you need plenty of contractors, installers, and people who are otherwise handy around the house.

Going this route could have a nice payoff. There are around 77 million freestanding houses in the United States,[20] and the average one releases roughly twice the amount of greenhouse gas as the average car.[21] Yet so far, we're just inching along. According to Energy Star, about 40,000 homes have been upgraded through renovation.[22] Add that to the 940,000 new homes built to Energy Star standards, and that leaves about 76 million houses left to go. Commercial property is not zooming ahead either. By 2007, only 4,000 or so commercial buildings had actually earned the Energy Star label.[23]

What do critics say about this approach? There's an old adage that a neoliberal is a conservative who's been arrested (and suddenly realizes how important the rights of the accused are) and a neoconservative is a liberal who's been mugged (and suddenly recognizes the importance of police protection and keeping the streets safe). But another category of liberals suddenly acquiring conservative spots probably includes people who have had to deal with construction permits and building codes. For some critics, strengthening building codes just means giving government (local, state, or federal) the chance to add more regulation, paperwork, permits, and red tape. In some places, building and renovation are already unbelievably complicated; in New York City, architects hire "expediters" just to go down to the buildings department to stand in line. Rather than adding new layers of regulation, critics say, the money and effort should go into educating people about the tax incentives and long-term savings on their energy bill.

ARE OUR HOUSES TOO BIG?

In the 1990s, the United States seemed to be in an age of supersizing—cars, housing, soft drinks, and waistlines all got bigger. It was the era of the gargantuan houses often dubbed "McMansions." What counts as a McMansion? There's no drive-through window required, but *New York Times* architecture critic Fred Bernstein described one as a place where the master bedroom is the size of a tennis court.[24] The infatuation with huge houses subsided over time, and the 2008 mortgage crisis and recession ended it (at least for a while).

We can't all live like the rich and famous, but over the last couple of decades a fair number of us have been aiming for our own mini-McMansions. Houses built since the 1990s have on average been 26 percent larger than existing ones, according to figures collected by the government.[25]

According to an analysis by the Energy Department, "Larger homes require more energy to provide heating, air-conditioning, and lighting, and they tend to include more energy-using appliances, such as televisions and laundry equipment."[26] If you have teenagers at home, you're probably following this line of reasoning easily—the bigger the house, the more lights there are to be left on. Yes, that's exactly the point.

Is it the moment for the "smaller is beautiful" American

home? More Americans do appear to be thinking along these lines, according to surveys by the National Association of Home Builders (NAHB). As recently as 2000, 51 percent of respondents said that they'd prefer more space with fewer amenities as opposed to a smaller house with "higher-quality products" and more amenities. By 2004, the number of people choosing the bigger house had dropped to 37 percent.[27]

Environmentalists would love to think this is mainly due to concerns about energy and global warming, and some of it may be, but population trends probably play a bigger role. In 1970, about 40 percent of households consisted of married couples with children. Now it's about a quarter.[28] The huge baby boom generation is aging. With the kids heading off to live on their own, a lot of boomers may be looking for smaller digs.[29] Even here, though, what's "big" and what's "small" may be in the eye of the beholder. According to another building industry study, homes in developments aimed at empty nesters are typically two-bedroom units with more than 2,000 square feet.[30]

What's more, zoning laws in some areas prohibit smaller houses and lots as well. Brooklyn Park, Maryland, residents who built small houses on narrow slices of land quickly found out that not all of their neighbors were charmed. For the owners, these tiny houses, ranging from 12 to 18 feet wide, were a smart use of space and more affordable than a larger house. But many of their neighbors feared crowding and declining property values.[31] Some Americans look at small houses and see them as sensible little energy savers. Others start thinking about Pete Seeger's infamous "little boxes made of ticky-tacky."

MAYBE IT'S WHAT'S INSIDE THAT COUNTS

George Carlin had a whole repertoire of subversive comedy routines, but one of the most subversive was about how all you really need in life is "a place to keep your stuff. That's all your house is," Carlin told us, "a place to keep your stuff. If you didn't have so much stuff, you wouldn't need a house. You could just walk around all the time." Carlin was on target there. Americans do tend to have a lot of stuff.

We have electric gadgets to dry our hair, brush our teeth, grind beans for coffee, carve the turkey, chill wine, and cook sandwiches on both sides at the same time, and we haven't started on the products from the home entertainment department. We have so many electric doodads that some of us even buy little kits to help us hide the wires

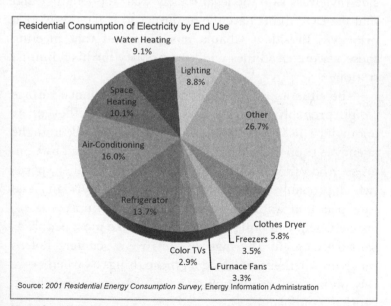

Residential Consumption of Electricity by End Use

Water Heating 9.1%
Lighting 8.8%
Space Heating 10.1%
Other 26.7%
Air-Conditioning 16.0%
Refrigerator 13.7%
Clothes Dryer 5.8%
Freezers 3.5%
Color TVs 2.9%
Furnace Fans 3.3%

Source: *2001 Residential Energy Consumption Survey,* Energy Information Administration

Most of our household electricity is used to either heat things up or keep them cold.

and cords. At least half the nation will breathe a sigh of re-
lief when the industry finally eradicates the whole dan-
gling, tangling cord thing.

The government's Energy Star program also sets standards for energy efficiency on everything from washers
and dryers to flat screen TVs, and overall, many appliances and electronics are more energy-efficient than they
used to be.

But here's the bad news: Even though Congress ordered
the Department of Energy to set energy standards for appliances after the 1973 oil embargo, the process has been
almost unbelievably slow. According to a report from the
Government Accountability Office, the Energy Department
missed all thirty-four deadlines for the standards it was
supposed to set. Worse yet, leaders in the DOE couldn't
even agree on why it was perpetually late. The GAO's outside reviewers said the legal review was "too lengthy" and
that the DOE didn't have the resources to make this work a
priority.[32] President Obama has promised that meeting
these various deadlines will be a priority for his administration.[33]

The situation with the Energy Star appliance ratings
would probably frustrate most Americans if they knew
about it. This is the law of the land, after all, and the
agency is supposed to carry out legislation passed by Congress. And other than the idea of living in smaller spaces
(which probably does run counter to what Americans tend
to expect and want), most of the ideas for improving energy efficiency in buildings probably strike most people as
sensible enough. The biggest controversy centers not so
much on whether we should do these things as whether we
rely on law and regulation or on education, outreach, incentives, and nagging.

However, all the caulking in the world won't take care

of all our energy problems. Improving energy efficiency in our buildings does help reduce greenhouse gas emissions, but it does almost nothing to reduce the country's reliance on foreign oil. Less than 2 percent of the country's electricity comes from oil.[34]

To reduce oil imports, we need to look at something some Americans may love even more than their homes—their cars. Scarlett O'Hara may have loved Tara even more than she loved Rhett Butler, but then again, Scarlett O'Hara didn't have a car.

Are We Looking to Live in All the Wrong Places?

If you want to start a fight at your next dinner party, you might consider introducing the topic of suburban sprawl. To its critics, sprawl peppers the countryside with strip malls and parking lots, it displaces farming and wildlife, and it means people drive miles to pick up a head of lettuce or buy a pair of socks. Fighting sprawl has become a movement. But for many other Americans, heading out of town means more afford-able housing, better schools, fewer crowds, and less traffic. What's more, the way they see it, the right to live where you want and use your own land the way you want to use it is fundamental. You can see why sprawl is such a controversial topic.

Another reason may be that Americans are just about split in half on where they say they'd like to live: Forty-eight percent say that if they could live anywhere, it would be in the city (23 percent) or the suburbs (25 percent); 51 percent want a less-crowded locale in a small town (30 percent) or rural area (21 percent).[35] Curbing suburban sprawl was originally viewed as a way to preserve farmland and protect wildlife. Now it's also one strategy to address the country's energy problems. To those who advocate this approach, curbing sprawl reduces commutes, conserves energy, and fights global warming.

ALL NEAT AND COMPACT

Putting new development close to existing population cen-ters and keeping communities more compact has all sorts of benefits, according to advocates. People will live closer to jobs, shopping, and other services because developers won't be able to build new houses way, way out of town. City government won't have to run water and power lines all over the county because new buildings will be located close to existing ones. Communities can protect green space because buildings won't be spread out over the whole countryside. And ideas such as rail and bus lines become realistic because you need a certain population density for any kind of mass transit to thrive. To some Americans, that sounds good. To others, it sounds like exactly what they've been hoping to avoid.

Personal preferences are important, as are people's concerns about affordable housing, good schools, and ease of transportation, but there's a more fundamental aspect to the dispute. By its very nature, land use planning limits where developers can build and what property owners can do with their own land. What rubs a lot of Americans the wrong way is the loss of freedom and control over land you've bought and paid for.

Oregon has been a pioneer in land use planning, and many consider it a model of "smart growth." Yet in 2005, six in ten Oregon voters approved Measure 37, requiring government to reimburse property owners for money they lose when they aren't allowed to sell their land to developers. Either the landowner is reimbursed or the development goes ahead.[36] For Measure 37's backers, the new law gives landowners the freedom to determine what they do with their own property. After all, backers argue, if you can't sell your land—or if you end up having to sell it for far less than what you might get from a developer—it isn't really yours anymore.

For advocates of land use policies, the Oregon law threatens everything they've worked for.*

BIG CITY, HERE I COME?

Curbing sprawl is one possibility, but some experts are going a step further. They believe that what we really need to do is get more people moving into cities. Former Milwaukee mayor John Norquist heads up the Congress for a New Urbanism, which considers traditional big cities such as Boston, Chicago, and New York "models of environmentalism." In cities such as these, more people walk, take public transportation, and live in apartments, which are generally more energy-efficient than freestanding houses.[37] Some experts have taken to battling the suburbs with a vengeance. James Howard Kunstler believes that "suburbia will come to be regarded as the greatest misallocation of resources in the history of the world." Kunstler fears that the United States will defend its "drive-in utopia long after it has become a terrible liability."[38]

Not all experts are so enamored of cities, though, especially very large ones. Some believe that they add to global warming because too much asphalt and concrete creates "heat islands." For these experts, the leafier suburbs, with lots of carbon dioxide-eating greenery, seem like the better option.[39] Then, there are the planners sketching out dreams of cities with plants and trees on every roof and maybe a windmill as well.

---

* You can check out the State of Oregon's summary of supporting and opposing views on Measure 37 at www.sos.state.or.us/elections/nov22004/guide/meas/m37_fav.html (supporting) and www.sos.state.or.us/elections/nov22004/guide/meas/m37_opp.html (opposing).

*   *   *

Back in the old West, disputes about land use between the ranchers and the farmers got so heated that some of them hired gunslingers to promote their point of view. That's how Billy the Kid got his start, and of course, Shane defended a homesteading family from a rancher's gang of thugs in the best western movie of all time.* Presumably we'll be able to resolve our differences now more peacefully than they did in the 1800s.

And we're not going to force people to move, the way Chairman Mao did when he ruled China. He sent well-educated urbanites he didn't like out to China's rural provinces for "reeducation." Even if all experts unanimously agreed that city living is better for the planet, no sane person would endorse that kind of Maoist social engineering here.

Still, in the future, we may not be plopping our new communities and developments down anywhere we want, the way we've often done in the past. There's likely to be more serious discussion about where we live, and a lot more talk about how to balance individual preferences with what's better for communities and the country as a whole.

---

* We are allowed to editorialize on non-energy-related topics.

On the Rebound:
Why More Efficiency
Doesn't Always Save Energy

By now, many Americans are chubbily aware of the false promise of the low-fat cookie. People leapt to the conclusion that "low-fat" meant "no consequences." But in fact what happened was that since the cookies seemed guilt-free ("It's low-fat! It won't hurt me!"), people tended to eat more of them than regular cookies, thereby keeping their waistline exactly the way it was, if not worse.

There's a similar phenomenon in energy, and it's one that explains a great deal about the country's energy habits and why we can't expect energy efficiency to solve all our energy problems. It even has a name—the rebound effect.

Back in 1865, a British economist named William Stanley Jevons noticed something weird going on in England's Industrial Revolution. The steam engine that drove the Victorian era's factories, ships, and railroads kept getting more efficient, so engines were able to do far more work with less coal. Yet coal use didn't drop—in fact, Britain was using more coal than ever before. Jevons concluded that the more efficient engine was actually encouraging people to use more coal. When steam engines became more efficient, they became cheaper to run, so people found new ways of using them—and also lost their incentive to conserve fuel. The better the engines got, the more coal Britain used.

This is what is now known as the rebound effect, or occasionally the Jevons paradox. Sometimes a new advance in efficiency can cut energy use when it's first introduced, but then people adapt and energy use climbs again. If you switch all the lightbulbs in your house from incandescent to compact fluorescents, your electric bill should go down. But because your bill has gone down, and since you know the fluorescent bulbs use less energy, you may not feel the same pressure to turn the lights out when you leave the room. So because the lights are on longer, your electricity use rebounds—maybe not to the same level it was before, but enough so you're not getting the full benefit of the efficient bulbs.

Exactly how the rebound effect works is one of the most furious debates in the arcane world of energy economics, for obvious reasons. Some people argue that this rebound effect undercuts the rationale for energy conservation. Essentially, they argue, why bother? No matter what you do, people will use more energy. Other economists argue that the rebound rarely cancels out the energy savings completely. You gain something from conservation, even if it's not everything you'd hoped.

Studies show the impact of the rebound effect depends on what's being made more efficient and whether or not you control the switch.[40] Refrigerators are more energy-efficient than they used to be, and there has been almost no rebound effect at all. Think about it: Would having a more efficient refrigerator make you more likely to leave the door hanging open? Probably not. But in other areas the rebound can be pretty steep: up to 50 percent for more efficient air-conditioning, 40 percent for water heaters, and 30 percent for space heating. Nobody wants to sweat any more than necessary, so if your air conditioner is cheaper to operate, you'll use it more.

There's also a rebound effect on auto fuel efficiency. After all, if you're driving an SUV when gas prices skyrocket, you may well drive less to save money. But if you trade in that SUV for a Prius, you're going to go back to your old driving habits. That was the whole point of changing cars: to do as much driving as you used to do at a price you could afford. Everyone agrees there's a rebound effect on cars; the big argument is about how much. The National Highway Traffic Safety Administration, which sets fuel economy standards in the United States, assumes a rebound of 10 to 20 percent, but other estimates go anywhere from 6 percent to 30 percent.[41]

Some experts contend these studies don't cover the full impact of the rebound. For example, maybe you'll take the money you save on more efficient appliances and spend it on even more appliances, like a second (or third) television or DVD player. Or maybe you'll get a more fuel-efficient car so you can afford a longer commute, buying a bigger house farther away from your job and using more energy on all fronts. All this is very hard to measure.[42]

Most experts agree on two things about the rebound effect. One is that you can't count on getting the full benefit of more efficient technology. In other words, doubling fuel economy for cars doesn't mean you're going to cut gasoline use in half. And second, greater efficiency alone isn't going to solve our energy problems. If we're really serious about cutting energy use, we're going to have to look at other ways of changing behavior, such as imposing higher taxes or altering social norms so that people who use energy lavishly really are good and ashamed of themselves.

CHAPTER 12

DRIVEN TO DISTRACTION

Americans are broad-minded people. They'll accept
the fact that a person can be an alcoholic, a dope
fiend, a wife beater, and even a newspaperman, but
if a man doesn't drive, there is something wrong
with him.

—*Art Buchwald*[1]

The automobile is central to American life—but you already knew that. Few things have been more thoroughly analyzed than the American car culture. Henry Ford developed the assembly line to build cars. The first rock and roll song was about a car (specifically, an Oldsmobile).* We could go on and on, but many others have already done that for us. All we'd suggest is that you think about your life for a moment. How much can you get done without a car? If you're like most Americans (in other words, not a New Yorker), the answer is "not much."

---

* There are multiple contenders for the title of "first rock song," but the Rock and Roll Hall of Fame has given it to the relatively obscure song, "Rocket 88," sung in 1951 by the Kings of Rhythm, including a very young Ike Turner. See www.rockhall.com/inductee/ike-and-tina-turner; www.time.com/time/arts/article/0,8599,661084,00.html.

Fewer than 9 percent of American households get by without a car, and six in ten households have more than one.[2] That helps make the car the most significant part of the energy debate for most people and the best example of the "triple threat" we've been talking about. You pay directly for the gas you put in your car, so the economic cost is right in your face. The oil we import is mostly going in your gas tank, so cars are a huge factor in our energy security. Lastly, auto exhaust is a major source of carbon dioxide, so it affects global warming.

No wonder, then, that the fight over energy focuses on cars—in fact, maybe it even focuses on them too much. There's just so much room for pontificating and finger-pointing on all sides. As we've pointed out before, only about a quarter of our energy use is for transportation, and of course not all of that is for passenger cars. There's no question, however, that we can't get a grip on our energy problems without doing something radical about how we travel. So the time has come. We have got to revamp our cars in some important ways, and in this chapter, we're going to take a look at our cars and what options we have for retooling them.

BIGGER, BETTER, AND FASTER OUT OF THE STARTING GATE

Raising fuel efficiency standards for cars is one of the most direct ways for dealing with this issue. In 2007, the federal government increased fuel standards for the first time in twenty years, and in 2009 the Obama administration announced the requirements would be raised even further, to 35.5 miles per gallon by 2016, an increase of 8 miles per gallon.[3] The overall fuel economy of American cars has been pretty much flat for twenty years. But if you've bothered to look under the hood at all, you know that automotive technology hasn't stagnated over that time. So why haven't we seen that in gas mileage already? Why aren't our cars more efficient?

The short answer is that they are; the internal combustion engine has improved dramatically over the last two decades. If anything, this has been a period of great engineering advances, some of which you've heard of, and most of which you probably haven't. Once automakers started putting in onboard computers to monitor engine performance, a whole host of advances became possible, such as technologies that turn off engine cylinders that aren't needed or variable valve timing. All of them add up to more power with less fuel.

But once you've made the engine more efficient, the question is, what do you do with that efficiency? And the answer is that we poured that gain into making vehicles bigger and faster. We didn't use it for better mileage.

Government statistics show the average light-duty vehicle built in 2007 (that covers cars, vans, pickups, and SUVs) weighed in at 4,144 pounds, nearly 900 pounds heavier than in 1987. (Must be all those cupholders.) Over the same period, horsepower has nearly doubled, from 118 to 223, and the average vehicle goes from zero to 60 mph in 9.6 seconds, three and half seconds faster than before. But mileage per gallon has actually gotten slightly worse, down to 20.2 mpg compared to 22 mpg in 1987.[4]

And yes, the drop in average fuel efficiency is mostly due to SUVs. Look at the statistics:

- Sales of pickup trucks have hardly changed since 1975. They account for about 13–14 percent of vehicle sales then and now.
- SUV sales, by contrast, grew from 1.8 percent of sales in 1975 and 6.3 percent in 1988 to nearly 29 percent in 2007.

- Proportionally, cars fell from 71 percent of all sales in 1975 to 45 percent in 2007, and trucks of all kinds (including pickups and SUVs) accounted for 49 percent of vehicle sales in 2007, compared to less than 20 percent three decades before.[5]

A SMALL DEFENSE OF DETROIT

Detroit's "Big Three" automakers made an enormous number of mistakes in the past few years that left GM and Chrysler no option but government bailouts and bankruptcy court. Lots of people count their love affair with the SUV as one of them. More than a few critics cite GM's decision to build the Hummer while dismissing hybrids and shutting down its innovative electric car project, the EV1, as a fatal error. In GM's defense, however, the situation looked quite different in the 1990s. Back then, SUVs looked like Detroit's salvation. The profit margins on SUVs were huge, more than $9,000 a vehicle by some estimates, and it was a market where American automakers led and foreign makers were playing catch-up.[6] By contrast, the EV1 was expensive and deeply unprofitable, particularly when fuel prices were low. In fact, while the EV1 had devoted fans (see the documentary *Who Killed the Electric Car?* sometime), it also had major critics. Despite its innovative air, *Time* magazine named it as one of the fifty worst cars of all time.*

---

* "The 50 Worst Cars of All Time," www.time.com/time/specials/2007/article/0,28804,1658545_1658544_1658535,00.html. To be fair, the Ford Explorer and the Hummer H3 are also on the list, precisely because they are egregious gas-guzzlers.

LIQUID ASSETS?

As with everything else about energy, there are plenty of ideas whipping around about ways to save the automobile, but not a whole lot of context. Fortunately, when you strip away all the politics and culture wars over why we aren't more like the Europeans versus who wants to be like the Europeans anyway, we face a simple choice: Do we want to continue putting something liquid in our tanks, or do we switch to electricity?

Liquid fuel, namely, gasoline and diesel, has been the option of choice for a century. Sure, in the early days of horseless carriages, there were other alternatives out there, like the Baker Electric and the Stanley Steamer. But they all fell by the wayside because gasoline engines were more powerful than the electric batteries of the day and much safer and easier to use than steam engines.

If we find a liquid alternative to gasoline, then not much about cars has to change: We still pump liquids into our tanks, which we'll continue to buy from the huge network of service stations that already exists. Automotive technology doesn't change very much, and we'd still end up with something cleaner. Biofuels, such as ethanol and biodiesel, fall into this category, as does compressed or liquefied natural gas.

Or we can switch to electricity. Technology has come a long way, and electric or partially electric vehicles are far more viable now. The electric vehicles themselves produce very little emissions (although you still may have emissions from generating the electricity in the first place). How you "fill up the tank" would be very different, since you'd need to plug the car into something and draw power from the grid.

Let's take a look at a few of the major options being

tossed around. Some are available right now; others are pretty far off.

Hybrids, Part I

The term *hybrid* is getting confusing, because it's used for several different things. So-called hybrid-electric vehicles, such as the Prius, have two engines: an electric motor for slow speeds and a conventional gasoline engine that kicks in at highway speeds. The gasoline engine also charges the electric battery. Hybrid-electrics still use gasoline, but much less than a conventional vehicle, and they perform the same on the highway, with a few minor differences. (For example, the gas engine doesn't idle. Since the electric motor starts immediately when the pedal is pushed, the car goes completely silent when stopped or at low speeds, which is a little freaky the first time you experience it.) That's why hybrid-electrics outperform conventional cars in city driving, where they're primarily running on their electric motor. In highway driving, there's generally a smaller gap between hybrids and conventional cars.*

Hybrid-electrics have some genuine advantages. You can use the technology in a wide range of cars, and automakers are already making hybrid luxury cars, pickups, and SUVs. You can walk into a showroom and drive out with a Prius or a hybrid Civic today and get first-class gas mileage: 46 miles to the gallon for a 2009 Prius (compared to, for example, 16 mpg for a Hummer).[7] Also, hybrid-electrics fit in perfectly with the automotive infrastructure we have now: You don't need new kinds of fueling stations, and you can get them repaired at any dealership.

---

* A hybrid 2009 Toyota Camry, for example, gets 33 mpg in the city, compared to 19 mpg for a conventional Camry. On the highway, however, the hybrid gets 34 mpg to the conventional car's 28 mpg.

The real problem is price. There's no question they use less fuel than comparable models (a regular Honda Civic gets a combined city/highway rating of 29 mpg, while the hybrid model gets 42 mpg). But they're still more expensive to make than conventional vehicles, so the sticker price for a hybrid is higher. In the fall of 2008, a hybrid would cost you anywhere from $2,000 to $8,000 more than a conventional gas-only car, depending on the model.[8]

The federal government does offer tax incentives for fuel-efficient vehicles, which can knock up to several thousand dollars off the price. However, the federal credit is intended merely to prime the pump, so to speak, and phases out when an automaker sells more than 60,000 hybrids. Toyota and Honda, the major sellers of hybrids, have already passed that mark, so there aren't any more tax incentives for those models.[9]

For the average person, the question is whether you save enough on fuel to make up the difference in vehicle price. That depends in large part on the price of fuel. In the fall of 2008, with gas prices at $4 per gallon, both *Consumer Reports* and *Kiplinger's* concluded that consumers would come out ahead on the five-year ownership costs of a hybrid, at least on some models.[10] But the hybrid advantage evaporates if the price of gas falls.*

Hybrids, Part II: Return of the Plug-In

You may have heard something about the Chevy Volt, General Motors' new plug-in hybrid vehicle. If you haven't, it's not for want of trying, because GM desperately, desperately wants you to know about it. Other carmakers are also

---

* As of this writing, *Kiplinger*'s has a nifty hybrid cost calculator on its website that lets you run the numbers yourself: www.kiplinger .com/tools/hybrid_calculator/index.html.

working on these, but GM is likely to be first to market. The giant automaker has all but bet the company on this new car, set to be available in 2010. If it succeeds, not only might it reverse GM's fortunes, but it could also fundamentally change how we power transportation.

Like other hybrids, a plug-in hybrid has both an electric motor and a gasoline engine. The difference is that the electric battery can also be plugged in and recharged overnight while it's sitting in your garage. That sounds simple, but the difference is huge because now the battery can be the *primary* power source for your car, at both city and highway speeds.[11] The gas engine will still kick in, but much less often, which means the gas mileage can be dramatically better than even a regular hybrid.*

You don't have to worry about the battery running down and leaving you by the side of the road—what gearheads call "range anxiety." The key number, auto engineers say, is 40 miles. If a plug-in can go 40 miles without a recharge or using the gas engine, you've covered the average commuting distance for most drivers without using any gasoline at all. This is the number one advantage of the plug-in: you get the environmental benefits of an electric car with the performance of a conventional one.

There are huge technical challenges here, mostly about how to make an electric battery that's powerful enough to do the job without weighing a ton or more. Remember, unlike

---

* The Volt, it appears, will differ from this, because the gas engine will not fully recharge the battery when it kicks in. Only plugging it in will do that. See Edmunds.com, "No 'Revolting' the Volt," September 23, 2008 (www.edmunds.com/insideline/do/News/articleId= 132112), and Jim Motavalli, "The Highly Charged Chevy Volt Controversy," *New York Times* "Wheels" blog, October 1, 2008 (http://wheels .blogs.nytimes.com/2008/10/01/the-highly-charged-chevy-volt-controversy).

the battery of a conventional hybrid, the plug-in's battery is doing most of the work. There's also the question of whether these cars will be able to plug in to household outlets or whether they'll need special adapters or even charging stations, much like gas stations.[12]

Also, you can't discount the economic challenges. At this writing, we don't know how much a plug-in hybrid is going to cost, but there's very little doubt that it's going to be more than a conventional car, and more than a hybrid-electric as well. At least to start with, you're going to pay a premium for this technology. So the trade-offs, in the end, are much like those of a regular hybrid. Whether that financial premium pays off for you personally over time depends partly on the price of gas. Of course, the social advantages of reducing emissions and oil imports are also part of the equation.

A plug-in hybrid is different from a pure electric car in that the latter has no gas engine at all. The EV1 was in this category, and so is the Tesla Roadster, the all-electric sports car built by a Silicon Valley start-up. Tesla Motors decided to roll with the fact that electric cars cost more and came up with a $100,000 sports car model, figuring that the luxury market was the best way to make the technology viable. The Roadster has gotten good

There are a lot of new options for powering cars, but making them practical and affordable is the real challenge. Photo by Charles Bensinger and Renewable Energy Partners of New Mexico. Courtesy of the U.S. Department of Energy, National Renewable Energy Lab

reviews from the car magazines, but as of this writing only a handful have been sold.[13]

Many energy and automotive experts are extremely charged up, if you'll forgive the expression, about moving to electricity for cars. But, as with everything else in the energy business, if this takes off, it's going to take time. At a recent conference, experts predicted it would be another ten years before a million cars are on the road, connected to a "smarter" electric grid.[14] And of course, there are some 250 million vehicles on the road in the United States right now.

Fuel Cells

Fuel cells got their share of attention a few years ago, when the Bush administration decided they should be a major focus of federal research and development.[15] Fuel cells powered the Apollo moon missions back in the 1960s and 1970s, and they've got numerous potential advantages today, not least of which is that they produce no emissions. And while they're most often talked about as a vehicle technology, they're got lots of other applications. Fuel cells would make great backup generators for hospitals, for example, and could potentially become a power source for ships or trains.

Essentially, a fuel cell is a form of battery, which makes electricity as part of a chemical reaction when two substances (usually hydrogen and oxygen) are mixed through an electrolyte. Unlike a regular battery, a fuel cell doesn't have to be recharged, although you do have to keep feeding it chemicals. The only by-product of the whole process is water, produced during the chemical reaction, which is no big deal. Fuel cells are potentially much more efficient than any combustion engine. And since hydrogen is the most common element in the world (number 1 on the periodic table, if you remember your middle-school science) and can be made with any energy source, we're not going to run out.

Right now fuel cells are expensive and heavy, which makes them difficult for automobiles. As of 2007, there were a grand total of 223 hydrogen-powered vehicles in the United States. That's right: 223, out of 250 million vehicles.[16] It's very likely that you'll see fuel cells used in applications such as powering buildings or large vehicles (for example, ships) before you'll be able to buy a fuel-cell car.* In fact, the Obama administration decided to cut off federal research money for fuel-cell cars in 2009. Why? Because other technologies are much closer to production.[17]

Biofuels

Biofuels, fundamentally, are a lot like Hamburger Helper. Just as Hamburger Helper stretches meat by mixing it with whatever "helper" is, fuels such as ethanol and biodiesel make oil stretch by mixing with it with grain alcohol.†

For many people, this has a beautiful simplicity. The Middle East may have oil, but the Midwest has grain, and lots of it. So if we can somehow grow fuel, we'll be golden. Plus, this technology is available now. In fact, if you live in the Midwest, the chances are good that you filled up your tank with a form of ethanol this morning. The benefits are many: For starters, based on tailpipe emissions, ethanol also produces smaller amounts of greenhouse gases than pure gasoline. Also, cthanol is about as bipartisan as energy policy gets, with President Bush raising targets for its use

---

* There's a lot more about how fuel cells work and how they can be used at the Energy Department's Energy Efficiency and Renewable Energy website, www1.eere.energy.gov/hydrogenandfuelcells.

† We've always been afraid to look at the label and see what's actually in there.

and President Obama calling for increasing them even further. The Energy Information Administration projects that the vast majority of the increase in renewable energy over the next few decades will come from increased ethanol use, and officially, the United States is committed to raising ethanol use from 7 billion gallons in 2007 to 36 billion by 2022.[18] So it's fair to say that the United States is betting big on ethanol, as are lots of other countries.

In the United States, when you're talking about ethanol, you're talking about corn. In 2007–2008, fully 25 percent of the corn grown in the United States was used for fuel.[19] In Brazil and some other countries, ethanol is made out of sugarcane; you can also use wheat, potatoes, or other vegetables. (If you think this sounds suspiciously like the still that Granny had going in *The Beverly Hillbillies*, you're not that far off, chemically speaking.) Ethanol comes in varying grades, depending on how much ethanol is mixed in with the gasoline. Almost all the ethanol used in the United States is E10, meaning 10 percent ethanol and 90 percent gas. You can put E10 into a regular car and it'll work fine. There's another version called E85, which is 85 percent ethanol and 15 percent gas. To use this, you need what's called a "flex-fuel" vehicle that will run on either gas or E85. The major automakers do sell them, but you'll be mostly filling it with gas. E85 accounts for only 1 percent of all the ethanol sold in the United States, and as of March 2008 there were only 1,365 gas stations in the country that sell it (out of 120,000 stations nationwide).[20] Half of those are in the five states that produce the most ethanol.*

Yet for all that, ethanol is still a very small part of our fuel use. The 7 billion gallons of ethanol we used in 2007 sounds huge, but it accounted for only 5 percent of the 140

---

* Minnesota, Illinois, Iowa, South Dakota, and Nebraska.

billion gallons of liquid fuel Americans consumed that year. In addition, the Energy Department itself says there's serious doubt whether we can actually reach the 36 billion-gallon target by 2022. How come? Well, for one thing, ethanol is expensive to produce, and the only reason it's competitive at all is because the government has been subsidizing it heavily for years. The federal government started backing ethanol in the energy crisis of the 1970s, with current subsidies amounting to about 52 cents a gallon. This has actually been one of the most consistent parts of federal energy policy, and one of the most frequently criticized.

Most forms of renewable energy are more expensive than fossil fuels and need subsidies. But ethanol is frequently mentioned as a particularly egregious form of "corporate welfare." The fact that ethanol producers include giant agribusinesses such as Archer Daniels Midland rubs many the wrong way. And the fact that the government has been so consistent in supporting ethanol may have more to do with the fact that Iowa holds the first presidential caucuses than with its merits. All presidential candidates love ethanol in January of an election year.

The other, very serious criticism is that corn may be inherently the wrong stuff to use for energy. This is one of the most furious controversies in energy policy, and one of the wonkiest. But we'll try to play Mr. Science here. First off, corn ethanol doesn't have as much energy "kick" as gasoline, or even other kinds of ethanol. Every kind of fuel can be looked at in terms of how much energy it produces per unit. With liquid fuels, the most common way is to consider Btu per gallon. A gallon of regular gasoline produces 115,400 Btu (there's a reason it's been the fuel of choice for decades). A gallon of E85 only provides 81,630 Btu, or about three-quarters as much energy as a gallon of regular gas.[21] This in itself doesn't mean you can't use ethanol for fuel,

ENERGY PROPERTIES OF SELECTED FUELS

	Gasoline	One gallon B20 biodiesel	One gallon E85 ethanol blend	One gallon liquefied natural gas	One pound compressed natural gas
Energy value compared to a gallon of gasoline	100%	109% compared to gas; 99% compared to diesel	72% to 77%	64%	17.5%

Source: Alternative Fuels and Advanced Vehicles Data Center, U.S. Department of Energy, http://www.afdc.energy.gov/afdc/fuels/properties.html

but it does mean you've got to use more gallons of ethanol to go the same distance as with gasoline.

Further, there's ferocious debate among scientists and environmentalists over whether corn ethanol takes more energy to produce than it actually provides. Obviously, anything that takes more energy to make than it offers when you put it in your car is a loser. There have been dueling studies on this for years now, and much of the debate centers on what gets factored into the calculations (the energy used to grow the corn, process the ethanol, and transport it to refineries, for example). Right now, it looks like corn ethanol is coming out ahead, with one study estimating that ethanol produces 25 percent more energy than it takes to create.[22] That said, this calculation depends heavily on whether you're using best practices to make the stuff. If you use inefficient techniques, then corn ethanol might well be an energy loser.[23]

There's also a fierce debate over whether biofuels actually cut greenhouse gases. Most experts accept that ethanol puts out smaller amounts of greenhouse gases at the tailpipe than petroleum, but some argue that when you factor in all the gases produced during its production and the deforestation that may result if countries such as Brazil clear away land for biofuels, ethanol may be a net loser there as well.

Corn looks particularly bad as a fuel if you compare it to sugarcane ethanol, the kind used in Brazil. Anyone who loaded up on Cocoa Puffs as a child won't be surprised to hear that sugar ethanol produces *eight times* as much energy as it takes to make. Nobody even questions whether you come out ahead on that score.[24] Sugarcane also produces more ethanol per acre than corn. So why not just use sugarcane? For one thing, it's more expensive. The United States grows lots of corn, and it's cheap to use for ethanol. While we do grow some sugarcane in the United States, sugar prices are higher than corn prices, and it would cost twice as much to use U.S. sugar for ethanol as corn.[25] By contrast, Brazil is the world's largest sugar exporter, meaning that it's much more cost-effective for the Brazilians. We could import sugar ethanol from Brazil, but the federal government has actually imposed a tariff to keep Brazilian ethanol out of the country, in order to protect the domestic industry.

Plus, it's going to be pretty much impossible to grow enough corn to meet our fuel needs. Think back to those ethanol stats we mentioned a few paragraphs ago, and think of it as one of those word problems you had to deal with in middle school. If the United States currently devotes 25 percent of its corn crop to ethanol, and ethanol is 5 percent of American gasoline use, how much of the corn crop would be needed to produce 20 percent of gasoline use? That's right: all of it. Which would significantly impact American production of corn dogs, not to mention everything else we use corn for.

In fact, that's optimistic: One study concluded that devoting *all* the nation's corn production plus *all* the soybean production to biofuels would only meet 12 percent of gasoline demand and 6 percent of diesel demand.[26]

That's why most of the advanced research on biofuels focuses on "cellulosic ethanol," which is a technical term for "plants people don't eat." This means waste products such as wood chips, fast-growing trees, and plants such as switch-grass, a plant that commonly grows on the prairie. Unlike corn, switchgrass is a low-maintenance crop that grows on land that frequently isn't that good for other crops.

Unfortunately, while there's been research into cellulosic ethanol since the 1970s, no one is actually producing it in the United States. The national goal for biofuels assumes that we're going to be using 16 billion gallons of cellulosic ethanol by 2022, and in early 2009 the Energy Information Administration bluntly said there's just no way to meet that goal, based on the current pace of research.[27]

Natural Gas

T. Boone Pickens made his billions in oil, but right now the only thing he likes better than wind power is natural gas. Specifically, liquefied natural gas, which can be used to power cars. It's right at the center of the Pickens Plan, his approach to reducing dependence on foreign oil. Essentially, Pickens argues that we should invest big in wind power for electricity, and shift natural gas use from power plants to powering cars. That would result in cars that emit much smaller amounts of greenhouse gases and also don't need foreign oil.*

The idea looked pretty good in the fall of 2008, when gas prices were $4 per gallon, and nowhere did it look better than in Utah. There's a lot of natural gas production in and around the state, and Questar Gas, a local utility company,

---

* You can find out more about the Pickens plan at www.pickensplan .com.

makes compressed-gas pumps at its facilities available to the public. All this means that while gas was $4 per gallon, compressed natural gas was only 87 cents—and you can't argue with a bargain like that. Or can you? To make natural gas work, you need a pump and a car that will use it, and right now both aren't so easy to find.

It worked in Utah, of course. Utahans rushed to grab every natural gas vehicle they could get, and bought conversion kits for more. But the fact is that Utah is one of the few places in the country where you can find a natural gas pump easily. As for the cars, there's only one car in America that is built to run on natural gas: the Honda Civic GX, which the EPA calls "the cleanest internal-combustion vehicle on Earth."[28] But it's only available in certain parts of the country, it's from $5,000 to $10,000 more expensive than a standard Civic, and you'll have to wait sixty to ninety days to get one (Honda sells so few that they custom-build each one).[29] You can also buy a conversion kit for a regular car, but it will set you back $12,000 or so. As a result of these constraints, there were only 116,131 compressed-gas vehicles in the United States in 2006, the last date for which numbers are available.[30]

Advocates say that could change easily. We already use natural gas widely, so the infrastructure could be developed. There are a few other issues, however. Like ethanol, compressed gas doesn't have as much energy punch as gasoline; in fact, it has significantly less (a pound of compressed gas has only 17.5 percent of the energy of a gallon of gas).[31] So that "range anxiety" starts to kick in again. The Civic GX gets 250 miles to a tank, half of what you'd expect with a conventional Civic.[32]

There's also the question of whether the United States would be self-sufficient in natural gas if we started using it for cars. That's a tall order in the short term since natural gas already supplies a fifth of our electricity. That said, natural gas is closer on the horizon than some of the other alternatives. If gasoline prices go back up and stay up, you'll be hearing more about this option.

"NICE RIDE"

There are other possibilities out there. The world is full of tinkering geniuses and thoughtful engineers, and a decent share of them have devoted themselves to cars. All we've done here is sum up a few of the major ideas floating around. But there are some guiding principles that will be relevant for any idea that springs up, now or later.

For one thing, somebody's going to have to be able to make money on these cars, however they run. The United States is not going to turn into the old Soviet Union, where a car is something you get because of services rendered to the Party. (Nor would we want to, for many reasons, not least of which was that the Soviets made terrible cars.) A car is still going to be something that people will need, and buy, for their own reasons, and someone else will have to find a way to build and sell them at a profit.

When they make those buying decisions, a few people will buy the most economical, environmentally friendly car they can, because that's what gets their wheels spinning. A few others will buy the gaudiest, most enormous trophy car they can, because that's what makes them feel good. And the vast majority will want something somewhere in between on that giant scale between enthusiasm and practicality. Most Americans want to protect the environment and not spend too much, and do it all while still having something that will cause their next-door neighbor to amble over, breathe in that new-car smell, and say, "Nice ride."

The Big Bang Theory of Auto Safety

So far at least, the most fuel-efficient cars are smaller, and bigger cars guzzle more gas. That's a reality behind one argument against fuel economy that resonates with many average drivers, although it actually doesn't get the respect it deserves from the experts. The line of reasoning goes like this: small cars may be more fuel-efficient, but big cars are safer. Sure, everyone wants to get more miles to the gallon. But as a matter of basic physics, when a big thing runs into a smaller thing, the smaller thing tends to lose. No one wants to have a flash of buyer's remorse about their Mini Cooper when a Ford Explorer is bearing down on them.

There's evidence for this, although if you took it to its logical conclusion, we'd all be doing our car buying at Ed's Used Monster Truck and Big Rig Emporium, and crushing [unoccupied] small cars under our wheels just for fun.* The good news, fortunately, is that auto safety experts say we don't have to choose between "big" and "dead."

When the National Research Council was asked to assess the impact of federal fuel economy rules in 2002, they concluded that one of the downsides was auto safety. In the 1970s and 1980s, manufacturers had made cars smaller and lighter to improve mileage, and the research council concluded that led to an additional 1,300 to 2,600 traffic fatalities in 1993.[33]

The council itself pointed out that size and weight are only two of the factors in traffic safety. Other safety advances counteract changes in size: improved air bags, antilock brakes,

---

* It *would* be fun; we're not denying it. We just can't endorse the policy implications.

and daytime running lights, for a start. There have also been many social changes as well. Almost every state has mandatory seat-belt and child-restraint laws, and attitudes about drunk driving have changed dramatically. In the TV series *Mad Men*, set in the early 1960s, there are scenes of parents letting their children crawl around inside a moving car, and of people shouting pointers to a drunk trying to start his car ("Your lights, Roger! Don't forget your lights!"). There's more truth in that picture of a cavalier attitude about driving than your parents or grandparents might care to admit.

Because of all this, driving is safer across the board, regardless of what you're driving or what hits you. There were 44,525 traffic deaths in 1975, before fuel economy standards were imposed, compared to 41,059 in 2007. That may not seem like a huge drop, but when you consider that the U.S. population has increased by more than 80 million people and we're driving more than twice as many miles per year, that's impressive. If you look at it by traffic deaths per miles traveled, we've cut the death rate by more than half (there were 3.2 fatalities for every 100 million vehicle miles traveled in 1975; by 2007, that was down to 1.37).[34]

The research council also reported that the safety impact of vehicle size depends on which vehicles get smaller. If you make SUVs lighter but keep passenger cars the same size, the people in cars are less likely to be hurt, and the people in the SUV may not be in any more danger, the council suggested.[35]

This idea has been incorporated in the latest changes to fuel economy rules, with standards set to encourage fuel economy in light trucks first, rather than across the board.[36] And of course, making vehicles lighter isn't the only way of making them more efficient.

In any case, it's worth remembering that, if you'll pardon

the expression, the biggest factor in traffic accidents isn't size. Yes, car passengers are more likely to be killed in multi-vehicle accidents, while SUVs are more likely to roll over because they've got a high center of gravity.[37] But the statistics also show, very conclusively, that the biggest cause of traffic accidents isn't what you drive, it's how you drive it. The National Highway Traffic Safety Administration says the major contributors to traffic deaths are alcohol (32 percent of all cases), speeding (31 percent), and not wearing a seat belt (a whopping 54 percent). And more than a few cases involve all three of these.[38]

Take Your Pick: Food or Fuel?

Sounds like an ugly choice, doesn't it? Yet that's what a number of developing nations said they faced in 2008, and they blamed the rise of biofuels.

Food prices spiked up worldwide in 2008, and while that was painful in the United States, it was flatly devastating in other countries. There were food riots in thirty nations; in Haiti thousands marched peacefully chanting, "We're hungry!"[39] At an international conference in June 2008, rising prices for energy and fertilizers got part of the blame. But many leaders of developing countries said the fact that major agricultural nations, including the United States, the countries in the European Union, and Brazil, were shifting cropland to biofuels was part of the problem.[40]

Several major international organizations backed them up. A World Bank report concluded that corn prices had risen 60 percent worldwide between 2005 and 2008, largely driven by U.S. farmers shifting corn production to ethanol. "The grain required to fill the tank of a sports utility vehicle with ethanol (240 kilograms of maize for 100 liters of ethanol) could feed one person for a year; this shows how food and fuel compete," the World Bank said.[41]

Other experts, however, say ethanol is only a secondary factor. The Congressional Budget Office concluded that ethanol accounted for 10 to 15 percent of the total increase in U.S. food prices between April 2007 and April 2008—a significant amount, but much less than other factors like the price of energy itself. [42]

The controversy is likely to continue. The Organization for Economic Cooperation and Development warned that world food prices may rise another 20 percent to 50 percent by 2016, and that biofuels would help drive that increase without providing enough environmental benefits to offset it.[43] For what it's worth, world food prices fell in late 2008, just like all other prices, as part of the global financial crisis.[44] But food experts think that biofuels will continue to be a factor in prices, at least until we can make them using plants people don't eat.

This is one of the fundamental issues with biofuels. They count as renewable energy, because you can always grow more crops to replace them. But the land you grow things on isn't

You can make corn bread and corn flakes with it, but it also makes ethanol. Now government researchers are looking at ways to use the stalks and husks instead of (or in addition to) the corn itself. Photo by Jim Yost. Courtesy of the Department of Energy, National Renewable Energy Laboratory

renewable; as the old joke goes about investing in real estate, it's valuable because the good Lord isn't making any more of it. So if a farmer is selling his corn crop for ethanol, he's not selling it for corn flakes.

This is what bothers some environmentalists about biofuels who argue that biofuels are the latest factor driving deforestation of the Amazon; the Brazilian boom in sugarcane ethanol is just another reason to clear away forests. Ironically, reducing the amount of forest (which absorbs carbon dioxide) makes global warming worse, not better.[45] There could be other environmental impacts as well: The National Research Council warns that using corn for ethanol could significantly impact water use and quality in the United States as new areas are opened up for biofuel production.[46]

We can, of course, duck this dilemma entirely by moving to "celluosic" ethanol, made from grass, wood, or something else inedible. But that technology is years away. In the meantime, we may need to ask ourselves whether we are actually getting all that much energy from the investment. As we mentioned earlier, devoting *all* the nation's corn production plus *all* the soybean production to biofuels would only meet 12 percent of gasoline demand and 6 percent of diesel demand.[47]

And then we'd still be hungry for more food *and* more energy.

CHAPTER 13

LOOKING FOR MR. WIZARD

We all agree that your theory is crazy, but is it crazy enough?

—Niels Bohr

Since 1790, the U.S. Patent Office has accepted more than 6 million patents for gadgets of all sorts, from the awe-inspiring to the annoying.[1] Patent number 5,184,830 is for the Nintendo Game Boy. We'll let you decide for yourself which category that goes in.[2]

Six million patents in a little over two hundred years is impressive even if a lot of them never come to much, and rightly or wrongly, Americans do tend to think of the United States as fertile ground for inventors. Starting in grade school, we're filled with stories about Thomas Edison, Jonas Salk, and George Washington Carver. The lightbulb, artificial heart, cotton gin, and sewing machine are just a few of the breakthroughs Americans have produced. (The Game Boy, by the way, was patented by a Japanese company, so there are plenty of inventive types elsewhere; more on that later.)

We'll need some scientific and technical breakthroughs to solve the country's energy problems, so the question for this chapter is whether the United States is doing enough to keep the advances coming.

CAN WE INVENT OUR WAY OUT OF OUR ENERGY PROBLEM?

You've probably already heard about some of the newfangled notions for generating energy in all sorts of unheard-of ways—wind-powered cargo tankers, getting energy from rice husks, and the like.[3] Some of the ideas are amazing, but it's important to stress that the United States isn't going to be able to invent our way out of our energy dilemma anytime soon. Counting on some extraordinarily brilliant someone somewhere coming up with the technological miracle that will solve all our energy problems—that's not a plan; that's like buying lottery tickets rather than saving for retirement.

Still, we'll definitely need help from our scientist and inventor friends to get over the hump—not to mention help from engineers, systems analysts, technology managers, and entrepreneurs. Here are just a few of the things we need some smart folks to work on:

- **Fuel from plants.** Secretary of Energy Steven Chu, a Nobel Prize–winning physicist himself, says that one challenge is finding "new types of plants that require little energy to grow and that can be converted to clean and cheap alternatives to fossil fuels."[4]
- **Asphalt and battery.** To some extent, the future of electric cars and solar and wind energy depends on someone (or more likely a bunch of someones) solving the battery problem. We need electric cars that can go long distances on small, cheap batteries or electric cars won't be practical and affordable. We also need batteries that can store wind and solar power effectively.

- **Scrubbing coal clean.** Then there's that whole idea of removing the carbon dioxide from coal so humans can use the Earth's massive supply of it without pushing global warming into overdrive. According to the big MIT study on coal, the basic knowledge to do it exists, but engineering the technology, fine-tuning the systems, figuring out how to store the stuff safely once you've gotten it out—all those things and more are still out there waiting to be done.[5]

What about all those ideas about getting fuel from algae or garbage or the ocean tides or even capturing the energy joggers create when they use the treadmill at the gym?[6] Some seem a little strange, but there are clever people thinking about them. Will they actually work in real life?

STEMMING THE TIDE

Missions such as these depend on having a healthy supply of smart, energetic science- and tech-minded folk—"STEM workers" is the phrase that's now in vogue (for "science, technology, engineering, and math"). Unfortunately, there are questions about whether the United States is developing enough STEM talent for the future, and it's not just finding the next Albert Einstein or Bill Gates. Progress requires talent up and down the line. Apart from Secretary Chu (because we just talked about him), you probably can't name too many professors who've won Nobel Prizes in physics and chemistry, but the country needs them all the same. We need professors for Caltech, MIT, and other big research universities. We need top-notch STEM people in corporations—energy companies, automakers, utilities. We need brainy people in government at the Nuclear Regulatory Commission, the Environmental Protection Agency, the Department of Energy, and elsewhere. The states and cities need this kind of talent too. And then we need a work-force that can build a zippier power grid and manage more complex utility systems. If we want buildings that are more energy-efficient, we need architects and builders who know how to do it. The list goes on, and we won't get through it unless we have a top-quality STEM talent pool.

Questions about STEM talent are usually raised in con-nection with the country's ability to compete in the global economy, but we need these people to address the energy/environmental challenges too. Do we have the people, or could we face a shortage?

Are We Training Enough STEM Professionals?

A 2005 study from the National Academy of Sciences, Na-tional Academy of Engineering, and the Institute of Medi-cine set off alarms about the country's supply of high-tech talent.[7] According to congressional testimony based on the study, fewer than a third of U.S. college graduates get de-grees in science and engineering fields—that's way behind what's happening in countries such as China, where 59 per-cent of students graduate in these fields, or Japan, where two-thirds do. What's more, many STEM doctorates go to students from abroad who may or may not stay in the United States after studying here (and who may or may not be allowed to stay under current immigration law). In 2003, nearly six in ten doctorates in engineering went to foreign students.[8]

Not everyone buys the idea that the United States has (or will have) a STEM worker shortage. Some say the cor-porate hue and cry over a "STEM worker shortage" comes mainly from corporations' unwillingness to pay salaries high enough to make the jobs attractive to Americans.[9] Yet even a severe critic such as Duke University's Vivek Wad-hwa (who once called the shortage idea "a lie") accepts that there could well be problems in specific areas.[10] Which ones could be trouble? Renewable energy and biofuels.[11]

Do American Students Learn Enough Math and Science?

The issue here is whether the public schools are turning out enough students with the knowledge, skills, and motivation to become the energy and environmental workers we're looking for. You don't go far in science, engineering, or even business unless you can handle advanced math. Yet, according to the National Assessment of Educational Progress (NAEP), only 23 percent of high school seniors are "proficient" in math.[12] Three of every ten college freshmen now have to take remedial math and science courses because they are not adequately prepared for college-level work.[13]

Should We Make It Easier for Foreign-Born STEM Professionals to Work Here?

In 2008, the country created sixty-five thousand visas for foreign workers with "highly specialized knowledge and a bachelor's degree or higher (or its equivalent), such as scientists, engineers, or computer programmers."[14] Whether that's too high or too low is the subject of a fierce debate that often echoes the country's larger debate over immigration. For many leaders of high-tech companies and academic institutions, the figure is way too low. In testimony before Congress, the president of the National Academy of Engineering, Dr. William Wulf, pointed out that a quarter of the engineering faculty in American universities and a third of U.S. Nobel Prize winners were born outside the country: "We have been skimming the best and brightest minds from across the globe, and prospering because of it."[15] *New York Times* columnist Thomas Friedman once recommended stapling a green card to the diploma of any foreign-born student receiving an advanced science or technology degree from an American university.[16]

Again, some opponents claim that the corporate interest is more about keeping costs down than worry over the country's future.[17] A 2006 GAO study confirmed that some companies do take advantage of foreign workers here on specialized visas (they're called H-1B visas), paying them less than they would have to pay a legal U.S. resident in the same job.[18]

ARE WE PUTTING ENOUGH MONEY INTO RESEARCH?

The government regularly invests in research and taxpayer dollars have laid the groundwork for breakthroughs such as MRI and CAT scans, cell phones, and satellite TV and radio.[19] So the question is whether the government spends enough in this area and whether this is the best way to push science and technology forward.

Since the 1970s, the United States has invested billions of dollars in energy R & D, much of it through the Department of Energy.[20] Over the years, the projects have covered the gamut, ranging from looking for ways to reduce the cost of building new nuclear plants to research on how to reduce mercury emissions—something actor Jeremy Piven might be happy to hear about. (If you haven't been following showbiz lately, Piven dropped out of a Broadway play due to mercury poisoning from eating fish twice a day, creating a huge controversy over whether he was really sick or just trying to get out of the show.)[21]

Given how little agreement there is on government spending overall, you'll hardly be surprised to learn that there are disputes on whether the government spends too little or too much on energy R & D. Many experts want the country to up its investment substantially, but not everyone believes there's a lot to show for the money. Some critics are also troubled by the idea of government "picking winners and losers" in science and technology, and others worry that members of Congress (who approve DOE's research allocations) mainly want to send projects to their own districts. As is often true, waste is in the eye of the beholder. People who like renewables often snarl at the amounts of money devoted to research on clean coal and nuclear energy. Others think renewables are getting more attention than they deserve. In the end, while almost no one thinks the U.S. government should get out of energy research entirely, almost no one is 100 percent happy with the way it's been done over the last decade or two.

IS IT TIME TO REBOOT?

Will the United States have the talent and research capacity to turn the corner on energy? Consider what the United States did after the Soviet Union launched Sputnik on October 4, 1957. The satellite didn't do much more than ping, but most Americans at the time saw the Soviet advance as a security threat, an economic threat, and a blow to U.S. pride. But here's what happened next.

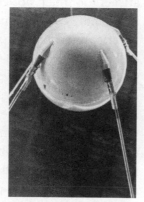

All the first Sputnik could do was send out a radio "beep," but after the Soviets launched the world's first satellite, the United States went into overdrive to improve science and technology capacities.
Courtesy of NASA

- By the end of November, President Dwight Eisenhower had created a presidential committee to give him advice on how to respond.[22]

- By July 1958, NASA had been founded.[23]
- A few months later, President Eisenhower signed legislation providing federal aid for science and math education at all levels.[24]
- Funding for the National Science Foundation went from $34 million to $500 million in the next decade.[25]
- By 1969, the United States landed human beings on the moon.

Of course, this all took place while the United States was waging a cold war with the Soviet Union. The question now is whether we can get our act together without having a "red menace" to spur us on.

TAKE THE MONEY AND RUN

Remember that patent for the Nintendo Game Boy? It didn't attract too much notice at the time, but it was an early clue that Japanese companies were about to beat the socks off American companies in consumer electronics. Are we about to let it happen again—this time in the competition to develop the new products that will solve our own and the world's energy problems? There are already signs that we may not be where we want to be. Most solar panels and wind turbines are imported from abroad, not manufactured here.[26] A recent "top 10" list in the fast-changing business of photovoltaic solar cells included companies from Germany, Japan, Taiwan, and China; only one U.S. company (First Solar) even made the list.[27] The truth is that there are plenty of smart, energetic people in other countries who want in on the game. The question is whether we plan on watching them play it while we sit on the bench eating popcorn?

Green-Collar Jobs, You Say?
Name Some

It's a phrase that rolls trippingly off the tongue these days. Elected officials and candidates who used to promise voters a chicken in every pot are now more likely to assure us that our economy will prosper from millions of new green jobs. Typical Americans are optimistic too; 86 percent believe that investing in alternative energy will create lots of new jobs.[28]

As we'll see below, not everyone believes that green-collar jobs are the wave of the future, but first, let's get some of the mystery out of the way. What kind of work do people in green-collar jobs actually do?

Essentially, green-collar jobs are any jobs that are being created (or will be created) as the United States shifts away from an economy based mainly on fossil fuels to one with a much greater emphasis on renewables and energy efficiency. That's everything from inventors and engineers to workers who come around to put insulation in the attic and help you install your new, more energy-efficient furnace. In fact, one of the most appealing arguments for the green-collar economy is that the

Will Larry the Cable Guy become Larry the Solar Panel Installer? Photo by Paul Torcellini, Courtesy of the U.S. Department of Energy, National Renewable Energy Laboratory

work wouldn't be confined to "knowledge workers" or the service sector.

Want some specifics? Here are seven types of work that are often mentioned when experts talk about green jobs.

★ **WANTED: wind turbine engineer.** People who work for companies that manufacture wind turbines and solar panels, including engineers, designers, production line workers, sales, and management.

★ **WANTED: electric distribution mechanic.** Engineers and workers with the skills to design and rebuild the electric grid. The grid has more than 220,000 miles of high-voltage wires and another 5 million miles of regular wires, so that could mean lots of work for lots of people.[29]

★ **WANTED: corporate sustainability manager.** Corporate employees who design and manage their companies' sustainability programs. Major American companies from Coca-Cola to Wal-Mart are going to have to refashion the way they get and use energy, so it will take people to figure out how to do that and then make sure it gets done. Rona Fried, who heads a networking service called Sustainable-Business.com, says that "every industry is going green, so every single industry now has the opportunity to post green jobs."[30]

★ **WANTED: weatherization crew workers.** People who can design, build, and maintain buildings that use energy more efficiently. The Apollo Alliance estimates that 75 percent of

American buildings will be renovated or built from scratch in the next twenty years.[31]

★ **WANTED: solar panel installers**. There will always be do-it-yourselfers out there, and it's just as easy for the average Tim Allen wannabe to fall off the roof while installing a solar panel as while installing the Christmas lights. That said, most of us wouldn't have a clue about how to install a solar-thermal water heater or put photovoltaic cells on the roof, so we'll need people to help with that. There are now about 25,000 to 30,000 people who do this kind of work. The Solar Energy Industries Association (it is what it sounds like) says that number could jump to over 100,000 in five or six years.[32]

★ **WANTED: power systems design engineer.** Computer software specialists to design and maintain all the "smart" monitoring systems that will be needed for wind farms, nuclear reactors, the electric grid, "smart" buildings, congestion pricing systems, and so on. One career advice website points out that "coders with experience using large scale enterprise resource planning have an edge here, as well as developers familiar with open source and web 2.0 applications."[33] If you understand that, you may have a future here.

Many of these jobs demand advanced technical training and/or or engineering degrees, so the debate on whether the United States is set up to educate and coach enough people to take on these jobs is another one of those wrinkles that hasn't gotten much attention yet.

JOBS THAT CAN'T BE SENT OVERSEAS?

The biggest debate about green-collar jobs, though, is how fast and how far to push in that direction. Some experts want the country to push toward a new "green" economy as quickly as possible, and they paint a captivating picture of thriving new industries creating a multiplicity of new jobs, much the way the computer and Internet industries did in the 1980s and 1990s. Wesley Warren of the Natural Resources Defense Council believes that "by repowering our nation with clean energy, we will create millions of jobs that can't be sent overseas. By harnessing the energy of the sun and wind, we can refuel our nation and end our addiction to oil."[34]

Others see dangers in government policies to promote renewable energy and green-collar jobs over traditional fossil fuels and workers in the coal, oil, and natural gas industries. Columnist Robert Samuelson thinks the employment potential of green energy is "wildly exaggerated." If it's jobs you want, he says, it would be better to ramp up domestic oil and natural gas production. Samuelson points out that you would need to double the number of jobs in the wind and solar industries to match what you would get from just a five percent increase in employment in oil and gas.[35]

In the end, the choice may not be nearly as stark as it seems. Energy experts emphasize that we're going to be using fossil fuels for many years to come even if we go full steam ahead with renewables. Energy and utility companies now routinely work with fossil fuels and a whole range of alternatives. For many of them, it's not green-collar jobs versus fossil fuel jobs. It's both.

CHAPTER 14

SITTING ON TOP OF THE WORLD

Consequences, schmonsequences, as long as I'm rich!

—*Daffy Duck*

We've been tackling energy from an American perspective because, well, that's where we keep all our stuff. The same probably goes for you too. But we hope we've made the point that we're in trouble for global reasons, and it's not easy to pull up the drawbridge and pretend the rest of the world isn't out there. In fact, it's impossible.

If you're concerned about climate change, for example, there's a reason why it's called *global* warming. Even if Country A takes the problem seriously and starts making changes, that can be canceled out by Country B if it throws the fossil fuel equivalent of a wild frat party. (And yes, as far as much of the world is concerned, the United States has been Country B for the past ten years or so.)

The other big global issue is that everybody in the world wants energy, and projections are that world energy demand will double by 2050.[1] Since energy is traded in global markets, and since most of the experts say there's no way the United States can be truly energy independent anytime soon, we're going to be competing with other countries for the energy that's out there. It's just a question of degree.

As we've said before, this isn't a bad thing. It means that people around the world are getting ahead, expanding their economies, and living better. If you're worried about global poverty and the more than 2.5 billion people who live on less than $2 per day, this is even more important. Right now an estimated 1.6 billion people, about a quarter of the world's population, don't have access to electricity. In some parts of the world, such as rural Africa, fully 90 percent of the people get their energy the old-fashioned way: burning wood, peat, or something else they grow and throw on a fire.[2] These desperately poor parts of the world have a long list of needs, including clean water, medical care, and better access to education. If they're ever going to have a better life, however, at some point they're going to need energy. There's only so far you can go in life without an electrical outlet.

THERE IS NO TOMORROW

We're at a critical juncture here. Nothing happens fast in the energy world, with changes frequently taking years to implement. Once an energy decision gets made, you live with the consequences for decades. Just about every energy policy expert we've found says that the choices we make in the next few years are going to decide how this plays out. "If we don't cure cancer in 20 years, like it or not the world will stay the same," said Nathan Lewis, a professor at Caltech. "But with energy, we are in the middle of doing the biggest experiment that humans have ever done, and we get to do that experiment exactly once. And there is no tomorrow, because in 20 years that experiment will be cast in stone."[3]

Neither of these trends is any secret, and lots of other countries are already working on dealing with them. In this chapter, we're going to give you a heads-up on the critical international choices that will have to be made. We can and should do plenty at home to meet these challenges. But there are several critical foreign policy areas to watch, and all are on the Obama administration's agenda. If we don't get these right, our domestic decisions may not matter much.

Making a Deal in Copenhagen?

This is the next major international opportunity to talk about energy. In December 2009, world leaders will convene in Denmark for the United Nations Climate Conference to discuss what will follow the Kyoto Protocol, the current world effort to control greenhouse emissions.

The deal runs until 2012, but given how complicated this problem is, it's not too early to start haggling about what comes next. The Kyoto Protocol called for different targets, but overall wanted the industrialized nations to cut greenhouse gases an average of 5 percent from 1990 levels, measured over 2008 to 2012.[4] As you probably also know, the United States did not ratify the Kyoto treaty, with President George W. Bush saying mandatory emissions cuts would do too much harm to the U.S. economy. The Bush administration pushed for voluntary emissions restrictions instead. For its part, the U.S. Senate passed a resolution in 1997—by a 95-to-0 vote—specifying that the United States should not agree to any international agreement limiting greenhouse gases that would "result in serious harm to the U.S. economy."[5] In any case, many of the countries that did ratify still aren't on track to meet their goals by 2012.

There are going to be several big issues on the table in Copenhagen.[6] The negotiators will revisit the emissions targets for industrialized nations, but perhaps the biggest area of debate will be what the industrial nations owe to developing countries. Countries still climbing out of poverty will grab on to the cheapest, most efficient energy they can get, and that's not going to be "green" energy unless the richer nations help them. Much of the conference will be about identifying the actions, financing, and institutions that can help developing countries get clean energy.

Clearly this has huge implications for the United States. If we're going to help other countries work out their energy problems, and bring up green solutions, Copenhagen would be an excellent time to step up and say so. But how the United States approaches the Copenhagen conference may depend on how it handles another question.

Getting Out of the "Suicide Pact" with China

You probably didn't think we had any deal with China on energy, much less a suicide pact. And you'd be right. But this is the phrase lots of energy experts and foreign policy wonks use to describe what's going on.

Between the two of us, the United States and China account for nearly 40 percent of greenhouse gas emissions worldwide. We're also going to account for a huge amount of energy use over the next few decades. In both cases, the United States is the reigning champion, but China is coming on strong. In fact, China has now surpassed the United States as the main producer of greenhouse gases, although the United States has pumped out more of these gases in total, and still uses much more energy per person.[7]

Because of that, the energy discussion between the United States and China has devolved into this conversation, repeated over and over until everyone's sick of it.

The Americans: We're happy to change how we use energy and cut greenhouse gases. But China's already producing more carbon dioxide than we are, and it won't do any good for us to change if China doesn't. Plus, China's a major economic rival. Why should we add to our energy costs, become less competitive, and maybe lose jobs, if China doesn't do the same? You guys are building cheap, smoky coal plants like there's no tomorrow, and you know better. So, after you.

The Chinese: But the United States has produced most of the greenhouse gas that's already out there, and you're the world's biggest energy gluttons per person. Sounds like you want to hide the beer keg just as the second wave of guests starts showing up. Plus, we've got a billion people to lift out of poverty, thank you very much, and the world should cut us some slack as we do it. So, after *you*.

Thus, stalemate, and a brief two-nation summary of the rich-nations-versus-poor-nations debate that underlies a good share of the arguing over the Kyoto and Copenhagen conferences. That's why energy experts have described this as a suicide pact. Neither country will agree to an international deal on energy, and the world as a whole won't be able to make much progress without these two big players.

That's not to say the Chinese haven't done anything. China has made some serious steps in the past few years, including setting ambitious goals to improve their energy intensity and increase their use of renewable energy. They've also moved on energy efficiency, such as increasing standards for appliances and cars and beginning to implement a new national building code. Still, they haven't slowed down on those coal plants, and they're having trouble meeting the targets they've already set.[8]

But many foreign policy experts believe there's an opening now. Chinese leaders have been more vocal about the need to confront climate change, while the Obama administration has promised to make alternative energy a priority. The two countries could agree to cooperate on key problems, no matter what happens at the international climate talks. For example, the United States and China could collaborate on energy research into both cleaner use of coal and alternative energy, according to a special commission set up by the Pew Center on Global Climate Change and the Asia Society. The two countries could also cooperate on financing these projects, which may be even more critical.[9] It'd also be helpful if the two nations agreed not to take advantage of each other's moves on this front.[10]

Dealing with Difficult People:
The Mideast and Russia

One of the most famous lines about oil comes from a French politician, who after World War I called oil "the blood of the earth . . . the blood of victory . . . the blood of peace."[11] You could fill a library with all the books written about oil and foreign policy, so we won't add to that list (although we'll give you suggestions in the appendix online). But there's no question that the need for oil affects our national interests, and thus our foreign policy. It's enough to say that most of the world's remaining oil lies in parts of the world that are, shall we say, problematic. The Mideast is at the top of the list, with more than 60 percent of all the world's proved oil reserves. It's important to remember, however, that the issues aren't limited to one region. Both Russia's

Vladimir Putin and Venezuela's Hugo Chávez have a tendency to whip out their pipelines at moments of international tension.

So you'll find lots of people who argue that if we weren't so dependent on foreign oil, we wouldn't have to worry about these unstable or hostile oil producers. Or, alternatively, they argue that if we could bring peace to the Middle East, we wouldn't have to worry about disruptions to our energy supply.

Our response to that is, "Oh, is *that* all?" If you've been following our argument in this book, you'll know how difficult energy independence would be to achieve, and very likely it'll never happen at all. The Council on Foreign Relations, the voice of the "establishment" in foreign policy, said, "Voices that espouse 'energy independence' are doing the nation a disservice by focusing on a goal that is unachievable over the foreseeable future."[12]

As far as solving the problems of the Middle East, these have bested the top diplomatic minds of the past century. The past decade has been more difficult than usual, with the Israeli-Palestinian conflict remaining intractable, the war in Iraq, and America's image in the region taking a nosedive. Obviously, how the Obama administration deals with these problems is critical. Anything that can be done to calm the region down will probably work to our benefit, in energy terms. Things that make the region more unstable will make things worse.

There are lots of ways our oil supply could be disrupted, and it's important to make a distinction between *deliberate* and *inadvertent*. For most Americans, the vision is still of the Arab oil embargo of 1973, where OPEC deliberately cut off oil supplies to pressure the West into dropping its support for Israel. The gas lines of that era remain a vivid memory, but there's one important thing to remember: The OPEC countries didn't get what they wanted. In the end, U.S. policy didn't change.

The best recent lessons on this come from Europe, which gets most of its oil and natural gas from Russia, and the Russians are well aware of the leverage that gives them. For example, in July 2008, the supply of oil to the Czech Republic just happened to drop 40 percent the day after the Czechs signed a deal to work with the United States on a missile defense radar system (a touchy subject with the Russians). The Putin government never acknowledged that they had anything to do with this, although they did order more oil to be sent shortly thereafter. The Czechs, perhaps because they've had a long history with the Russians, didn't suffer too much, because they had prudently built an oil pipeline to Western Europe just in case. Again, it's important to note that while the Russians may have made their point, they didn't get what they wanted.[13]

There are lots of inadvertent things that could happen to the oil supply, however. Again, a recent incident involving Russia is a good example. In January 2009, Russia and Ukraine got into a dispute over natural gas prices that ended up cutting off the main pipeline transporting gas from Russia to Europe. After two weeks, the nations reached a deal, but not before millions of Europeans were left without heat, and, in some cases, without work as factories shut down.[14]

Sadly, this doesn't seem to work both ways. Often you'll hear people argue that by importing oil, we're sending money to nations that hate us or even support terrorists. There's truth in that. But they'll also argue that by cutting our oil imports we'll be starving these countries of funds, and that may not be true. As we've said, world demand for oil is rising, and if we buy less oil from a troublesome nation, it's likely someone else will buy more. It's illegal to import Iranian oil into the United States, since that country is on the State Department's official list of state sponsors of terrorism and because we're trying to prevent them from developing nuclear weapons. U.S. and international financial sanctions have certainly hurt the Iranian regime, making it hard for them to get the investment money they need to upgrade their aging oilfields.[15] But America's refusal to buy Iranian oil hasn't put them out of business yet. Iran has managed to cut oil and natural gas deals with China and Russia,[16] and is pursuing a natural gas pipeline to Pakistan and India,[17] all of whom are willing to buy their energy and help them invest.

The best you can say is that if we import less oil, at least it won't be our money going into the wrong hands. Which is something, but maybe not all we wanted.

Events such as these are why foreign policy experts don't like to talk about energy "independence," as we've noted; they like to talk about "energy security." The problem in Europe in 2009 wasn't just that they were using imported natural gas, it was that they were getting most of the gas from a single, famously irritable nation, across a single vulnerable pipeline. By contrast, the Czechs were able to ride out the 2008 oil shortage because they'd been prudent enough to have a backup plan.

That strategy, fundamentally, is the secret to energy security. You can import all the energy you want, and most of the countries on this planet will have to do that. But the energy has to come from stable sources, or at least you need a backup plan if anything goes wrong. Most historians argue that the 1970s OPEC embargo was a failure. And one major reason OPEC never tried it again is because lots of other countries entered the oil market, which means there are others to take up the slack should there be any new embargo.

The West got wise in the 1970s. The industrial world now has strategic oil reserves in case of any disruption, and the International Energy Agency has a global plan for coping with problems. If you remember the chart in Chapter 5 about where our oil comes from, you'll see that the United States actually doesn't depend as heavily on Middle Eastern suppliers as it does on nations such as Canada and Mexico. Yes, Saudi Arabia and others in that region are big players, and they can disrupt the market, but nobody can cut us off totally.

When it comes to energy security, as in so many things in life, it's worth pondering the lessons imparted by James Bond movies. So let the last word on this come from the venerable Desmond Llewelyn, hanging up his lab coat in *The World Is Not Enough* after seventeen movies as the gadget guru "Q." The movie plot actually turned on the politics surrounding a Russian oil pipeline, but the plot isn't much use to us in the real world. Rather, we're thinking of his advice to 007.

"I have always tried to teach you two things," Q tells Bond. "One, never let them see you bleed."

"And the second?" Bond asks with a raised eyebrow.

"Always have an escape plan," Q says as he vanishes down a trap door.

Setting the Global Thermostat: How to Tell if We're Really Making Progress on Climate Change

If you've been paying attention to the debate over greenhouse gases and global warming, you know that everybody has a target they want to set. Should we cut greenhouse gases back to 1990 levels by 2020? Was it cutting emissions 50 percent by 2025 or 25 percent by 2050? After a while it starts to blur.

However, there are fundamentally three factors that drive a country's greenhouse gas emissions. If you keep your eye on these three factors, it not only clarifies what needs to be done, it also shows you the enormous amount of work it's going to take to do it. So watch these numbers:

Population growth. The more people a country has, the more energy gets used, because there are more people driving around, using electricity, and so on.

Income per capita. When a country's economy grows, people have more money, and that means they can afford to use more energy. New businesses open up, more houses get built, more people get jobs and commute, people buy more stuff, and more energy gets used.

Intensity of emissions. This is more complicated. (And who knows? After *Quantum of Solace*, it might work as the title of the next James Bond movie.) Actually, it's a measure of how much greenhouse gas we're pumping out in proportion to the economy. Specifically, that means how many tons of greenhouse gases a country produces for every million dollars of gross domestic product.[18] You want this "intensity" number to be going down.

The good news is that the world in general and the United

States specifically are both getting more efficient and lowering their emissions intensity. The bad news is that we're not doing it fast enough to get ahead of the growth in population and the economy.

For the United States, the numbers work out like this:

Between 1990 and 2000, our average "intensity rate" improved as people became more efficient in using energy. But our population grew an average of 1.2 percent a year and the economy grew an average of 1.8 percent a year. So total greenhouse gas emissions went up 1.4 percent a year, despite the improvements in intensity.[19]

Since 2002, U.S. emissions intensity has fallen about 2 percent a year, and in 2006, our total greenhouse emissions actually declined. Partly that's because of utilities shifting to cleaner fuels, like natural gas and renewable energy.[20] But some of the improvement is thanks to good weather, which held down heating and cooling costs, and higher energy prices, which cut consumption—not exactly things we could plan. Those are also factors that can easily change—in 2007, greenhouse emissions bounced back up because of tougher weather that required more energy for both heating and cooling.[21]

This also gives you an idea of the challenges facing other countries. If it's any consolation (and if you're worried about global warming, it probably isn't), the chart on page 259 gives you an idea of how difficult it is to cut greenhouse gases worldwide. Some people argue that countries such as China are the new leaders in producing greenhouse gases, and why should we cut back when they don't? Well, it's not that the Chinese and other fast-developing countries such as India, Brazil, and South Korea are sitting on their hands. All of them are making solid strides in improving their energy intensity—

A GLOBAL EMISSIONS SCORECARD

Country	1990 CO_2 emissions (million tons)	2004 CO_2 emissions (million tons)	Percentage increase/decrease
United States	4,818	6,046	+25%
China	2,399	5,007	+109%
Russia	1,984	1,524	−23%
India	682	1,342	+97%
Japan	1,071	1,257	+17%
Germany	980	808	−18%
Canada	416	639	+54%
United Kingdom	579	587	+1%
France	364	373	+3%
Brazil	210	332	+58%
Spain	212	330	+56%
Ukraine	600	330	−45%
Saudi Arabia	255	308	+21%
Poland	348	307	−12%
Malaysia	55	177	+221%
World total	22,703	28,983	+28%

Growth in carbon dioxide emissions, selected countries, 1990–2004.
Source: UN Human Development Report 2007/2008, Table 1.1

but their rising populations and growing economies are making it hard to keep up. In the 1990s, China's intensity rate dropped by an average of 7.1 percent, to our comparatively measly 1.6 percent. But their population grew an average 1.1 percent a year and their economy grew a staggering 9.4 percent annually, wiping out any advances they made.

Worldwide, only a few countries have actually managed to reduce their greenhouse emissions, but nobody really wants to follow their example. During the 1990s, Russia and Ukraine saw average annual emissions drops of 3.9 percent and 6.3 percent, respectively.[22] Unfortunately, the main reason is that their economies totally tanked for the decade, to the point where people started emigrating elsewhere. Obviously, very few countries want to combat global warming by having their citizens flee the country or having their economies slide into a deep recession. The global financial crisis of 2008–2009 is already cutting greenhouse emissions. In the United States, carbon emissions dropped 2.8 percent in 2008, primarily because of the poor economy.[23] But you'd have to be Pollyanna on antidepressants to argue that subprime mortgages and reckless debt are good ways of setting environmental policy.

There are better role models. Britain and France have seen economic growth comparable to that of the United States while also controlling their greenhouse emissions.[24] One reason is that they, like most European countries, had much smaller population growth. But both have also made different choices on energy policy. In the 1980s, France invested heavily in nuclear power, and in the 1990s Britain began a major shift from coal to natural gas for electricity. As a result, both nations have essentially held their greenhouse emissions at about where they were in 1990.

So it doesn't take a recession to cut greenhouse emissions and energy use, but it does take a certain amount of determination. The main thing is that when politicians, energy executives, and environmentalists talk about what we should do about energy and global warming, we need to be certain that their proposals keep us ahead of the economic and population trends that naturally spur greater energy use. Because if the changes are too halfhearted, we'll end up running in place.

CHAPTER 15

SO NOW WHAT? IDEAS FROM THE LEFT, RIGHT, AND CENTER

If you have always done it that way, it is probably wrong.

—*Charles F. Kettering*

There may be fifty ways to leave your lover, at least according to Paul Simon, but there are hundreds, even thousands of ways the United States could begin solving its energy problems. At any one time, there are dozens of bills in Congress, and that's on top of hundreds of measures on tap in statehouses, city halls, community boards, and the like. Some of the proposals are straightforward and easy to understand; some are nearly incomprehensible (even the elected officials have to call in experts to help them understand them). One of the fifty tips from the Simon song is to "make a new plan, Stan." Planning is crucial for solving the country's energy problems. Pretending to go out to the store for a six-pack and never coming back isn't going to cut it.

Chapter 15 is a starter kit for coming up with your own energy plan for the United States. We'll summarize some important proposals experts and leaders are talking about. Since we're pretty sure you don't want to still be reading this book several months from now, we've omitted some good ideas. It's like those lists of the one hundred best books ever written or the one hundred funniest movies—some terrific stuff just doesn't make the cut.

NO ROLLERBLADING REQUIRED

We have tried to include proposals that tackle the energy problem from different angles and come from across the political spectrum. We've also emphasized proposals that seem relatively doable in the near term, as opposed to proposals that depend on a major scientific breakthrough or a massive cultural and political transformation—say, like everybody suddenly deciding to give up their cars and bicycling or Rollerblading to work. Take a look at the appendix online for some tips on finding more proposals.

If you've read every single word of this book up to now, you'll realize that we've touched on some of these ideas earlier, so this is more like a review and a chance to look at different proposals side by side. If you haven't read every single word (we consider skimming a perfectly honorable trade, by the way) and things get confusing at any point, you might want to head back to some of the previous chapters for more explanation.

As you work your way through the list, remember that the United States is going to have to do a variety of things. You can't just pick the one idea on the list that best suits your fancy and then forget about all the rest. You'll need proposals that address cars and transportation, housing and electricity, business and industry. We use energy in lots of different ways, and we need to rethink all of them.

Our last piece of advice is to try to be realistic about how quickly the country can get from where it is today to where you want it to be. Don't lose sight of the basics.

YOU MUST REMEMBER THIS

- The United States overwhelmingly relies on fossil fuels. In 2007, renewables supplied about 7 percent of America's energy, and wind and solar power were just a small sliver of that.[1]

- The United States produces about 10 percent of the world's oil, but we consume about 24 percent.[2]

- Oil (used mainly for transportation) produces about 44 percent of the country's greenhouse gas emissions.[3]

- Transportation and generating electricity are the major sources of energy-related greenhouse gas emissions. In 2006, transportation was responsible for 34 percent and electricity generation was responsible for nearly 40 percent.[4]

- According to the Department of Energy, the country's transportation needs will increase by 31 percent by 2030. The demand for electricity is expected to increase by 35 percent in the same period.[5]

- Demand for energy is expected to increase worldwide, maybe by as much as 45 percent by 2030. So even if we continue to import energy (and we probably will), there's going to be a lot more people out there trying to do the same.

So what are the big ideas? What kinds of solutions are experts and policymakers debating? Here are some to consider from the left, right, and center.

Raise taxes on gasoline. This is probably not where you want to start—not many people do. A gas tax could be the least popular idea out there, yet many experts say this is probably the most powerful, fastest-acting step we could take to get a grip on our energy problems. Americans' behavior over the past forty years gives gas tax advocates all the evidence they need. When gas is cheap, we buy big cars and drive them like there's no tomorrow. When it gets expensive, we trade in the behemoths for energy-saving models, consolidate our errands, and properly inflate our tires; some of us even take to biking. When gas prices plummeted in 2008, *New York Times* columnist Thomas Friedman was one of many experts who called for a gas tax. "How many times do we have to see this play before we admit that it always ends the same way?" Friedman asked.[6]

There are many ways a gas tax could work. One proposal is a sliding tax that would keep gas at, say, $3.50 a gallon no matter what the wholesale price of gas was.* Columnist Robert Samuelson suggested raising the gas tax by a penny a month for the next four years.[7] Some proposals suggest raising the gas tax but reducing payroll taxes to soften the blow for drivers.[8] Some suggest using gas tax money to develop renewable energy. What's behind all these variations on a theme? Experts see many benefits—reduced greenhouse gas emissions, keeping more dollars in the United States, reducing the power of big oil producers such as Russia and Saudi Arabia, and giving automakers and innovators an incentive to produce cars that use less gasoline

---

* For example, if gas cost $3, the tax would be 50 cents; if it cost $2, the tax would be $1.50. Right now gas taxes are set at fixed cents-per-gallon levels, regardless of what a gallon actually costs. As of January 2009, the federal tax was 18.4 cents per gallon, and state taxes averaged 21.59 cents. Of course, state taxes vary, ranging from 7.5 cents per gallon in Georgia to 37.5 cents in Washington state (www.eia.doe.gov/pub/oil_gas/petroleum/data_publications/petroleum_marketing_monthly/current/pdf/enote.pdf).

or none at all.[9] We probably don't have to spell the cons out for you. This will hurt nearly everyone. It's especially painful for people with long commutes and hardest on low-income people. It's hard to argue that it wouldn't push most of us to use less gas, but right now, survey data show the American people solidly against it: More than seven in ten reject the idea of mandating $4-per-gallon gas to encourage alternative energy. Nearly six in ten, 58 percent, say they are *strongly* opposed, which in survey terms is huge.[10] Plus, more than half oppose increasing gas taxes no matter how the tax was portrayed, whether as a boon to energy independence, an antidote for climate change, or a source of funds for repairing roads and bridges.[11] Politically, this is a tough sell.

Make it easier to drill for oil and natural gas in the United States. Unless you slept through the entire 2008 presidential election (which would have been very restful, albeit in Rip van Winkle territory), you know that some Americans want to "drill, baby, drill," and others don't. The Heritage Foundation, for example, argues that making it easier to tap domestic sources of oil and natural gas will "bring down energy costs for all Americans" and "make the nation less reliant on unfriendly or hostile nations."[12] Drilling advocates argue that limits on energy exploration and extraction offshore and in Alaska and other federal lands are outdated given the advances in drilling technology and the country's urgent need for more domestically produced oil. Columnist George Will points out that Hurricanes Katrina and Rita destroyed or damaged hundreds of drilling rigs without causing a large spill.[13] Heritage Foundation experts Nicolas Loris and Ben Lieberman propose that new drilling projects be required to use state-of-the-art technology with a proven track record for limiting the risk of spills.[14] Columnist Robert Samuelson makes the case that more domestic drilling development would "immediately boost high-wage jobs [such as] geologists, petroleum engineers, roustabouts."[15]

Environmental groups generally oppose drilling, saying that it endangers wilderness, wildlife, and the shoreline for little payoff.[16] Actually, their main argument could easily be summed up as "And for what?" They say that the additional energy would be minimal given the country's massive needs, and the little it would add would not be available for years. Since burning natural gas is less harmful to the environment than other fossil fuels, some Americans support drilling for it, but not for oil. Then there's the priority question. Out of all the things we could do, is this where we should put our time and money?

Organize a cap-and-trade system where companies buy (and sell) permits to release greenhouse gases. This one might be all wrapped up by the time you read this, or at least as wrapped up as things get in Washington. The House of Representatives passed a cap-and-trade bill in June 2009. Whether it becomes law or not, it's likely to be a contentious election issue for years to come. Political fireworks not withstanding, we're willing to bet that very few Americans understand what it is or why it's so controversial. It's based on a fundamental economic principle—the same one behind the gas tax. If something costs more, fewer people will do it. Therefore, if we want utilities and industries to rein in greenhouse gases, we need to make them pay more for doing it.

Cap-and-trade works something like this. The government sets a limit on how much greenhouse gas can be released in the United States. That's the cap. Then it gives (or sells or auctions) permits to each company to produce a certain amount of those gases. Since there are only so many permits, it's a little like rationing. If one company doesn't need to use its permit (maybe it relies on solar power), it can sell it to someone who does (such as a coal-fired power plant). That's the trade part.* The bottom line is that if you do something that releases significant quantities of greenhouse gases, you'll have to pay for it. For cap-and-trade supporters, a big plus is that the United States will actually set specific targets for reducing total greenhouse gas emissions and then organize the cap-and-trade system around them. What's more, the EPA used a system like this to reduce the emissions that cause acid rain, and the EPA says that it reduced them "faster and at far lower costs than anticipated."[18]

The downside? Some see the cap-and-trade approach as unnecessarily convoluted and perhaps even duplicitous. Columnist George Will called it "a huge tax hidden in a bureaucratic labyrinth of opaque permit transactions."[19] There are lots of disputes over how high or low the emissions cap should be and whether the government would give out the permits or sell them and at what cost. Some economists say a straightforward tax would work better (see page 268). But the major issue for most opponents is cost. This will make using coal, natural gas, and oil—the fossil fuels we rely on most—considerably more expensive. You don't really think these companies are going to absorb these extra costs themselves, do you? The Heritage Foundation argues that it will "raise energy prices, which disproportionately hurts low-income households," and undercut coal, "the one energy source America has in great abundance."[20] Europe already has a system like this for greenhouse gases,

---

* If you were ever perplexed by the idea of papal indulgences back in the Middle Ages (you could pay the church and get forgiven for your sins), this may sound familiar. Basically, if you come up with the money, you get a pass on doing something bad.

which is either succeeding in driving companies to cleaner renewable fuels or failing and creating a total mess, depending on whom you talk to.[21]

Tax energy from fossil fuels. Politicians often prefer cap-and-trade because it makes using fossil fuels more expensive but doesn't involve explicitly raising taxes (which politicians don't like to say they have done). However, many experts believe that a straightforward tax on energy produced from fossil fuels would be more effective and less complicated.* The Price Carbon Campaign, for example, has outlined a very specific plan for phasing in a carbon or fossil fuel tax over the next ten years. Their idea is to tax coal, oil, and natural gas at the point of origin (when coal is mined or when oil arrives at the port).[22] The tax would vary depending on how much carbon dioxide is released when the fuel is used (highest for coal and lowest for natural gas). Obviously, producers and importers would pass their costs along to utilities, refiners, and distributors, who would pass them on to consumers. According to the group's calculations, consumer costs would be about $180 the first year, growing to $1,800 annually a decade later, but they propose offsetting these costs by reducing payroll taxes or sending "carbon dividend" checks out to taxpayers.[23]

Clearly, there are many different ways of setting and collecting a carbon tax, but the general idea has some surprising supporters. Most are probably environmentalists. The Price Carbon Campaign is backed by Friends of the

---

* The GAO convened a panel of experts to look at the pros and cons of a tax versus a cap-and-trade approach, along with some hybrid plans. Their report to Congress is *Climate Change: Expert Opinion on the Economics of Policy Options to Address Climate Change*, May 2008, www.gao.gov/new.items/d08605.pdf.

Earth among others. But conservative columnist David Frum has called for a standby excise tax on oil (to be applied when the price drops below $65 a barrel), accompanied by cuts in the corporate income tax and/or other taxes.[24] Exxon CEO Rex Tillerson probably doesn't want a tax or cap-and-trade, but given the choice, he believes that a tax is "a more direct, a more transparent, and a more effective approach."[25] Plus, the nonpartisan and respected Congressional Budget Office concluded that an emissions tax would be the "most efficient" approach to reducing carbon emissions and "relatively easy to implement."[26]

The bottom line for many experts who've looked carefully at the long-term prospects for wind, solar, nuclear, and clean coal technology is their judgment that these ideas will never be competitive in the marketplace unless fossil fuels cost more than they do now.[27] (Skimmers, go back to Chapters 6 through 10 if we're losing you.) So for them, the question is whether to do it through a market system such as cap-and-trade or cut to the chase and just tax the stuff.

Here's another idea where the cons are pretty obvious. A carbon tax would make the energy most of us use regularly more expensive, and many opponents fear its impact on economic growth, especially when the economy is weak or recovering. And it's very likely that some parts of the country (the Midwest and coal country) would be much harder hit than others. Others have raised questions about whether penalizing carbon emissions (through a cap-and-trade approach or through a tax) would encourage more American companies to move their manufacturing operations to countries with fewer environmental regulations.[28]

Ban new coal-fired power plants (unless they have the latest clean coal technology). Since using coal to generate electricity creates over a third of the country's greenhouse gas emissions, some people think that making it more

expensive isn't going far enough.[29] They want to stop building any more coal-fired power plants unless they have a
new technology that removes carbon dioxide before it's released into the air (carbon capture and storage). Al Gore,
environmental groups such as the Sierra Club,[30] and James
Hansen, the NASA scientist who was one of the first to
warn about greenhouse gas emissions, want the country to
call a halt to building new coal plants that use coal the old-
fashioned way.[31] Since you can generate electricity with
nuclear, wind, solar, hydro, and geothermal power without
creating any greenhouse gases at all (or use natural gas and
emit a lot fewer of them), why would we keep using coal?
they ask. And if a utility really wants to use coal, it should
build in the new clean coal technology that removes carbon
dioxide from the process.

So what's the drawback? Money—and plenty of it. In
most areas of the country, coal is by far the cheapest way to
generate electricity. The new clean coal technology isn't just
expensive, it's still in rehearsal, so to speak. Banning the
old technology could bring the entire coal industry to a
near standstill for some period of years.

It would almost certainly hit some parts of the country
much harder than others. The Midwest gets much of its
electricity from coal, and the mining industry (which might
face falling demand for coal) is is a key employer in states
such as Kentucky, Pennsylvania, and Ohio. A variant of this
idea is to ban new coal plants unless they are designed so
they can be retrofitted with carbon capture and storage
technology when it is more broadly available, but this option is very costly as well.[32]

Fast-track the building of new nuclear power plants.
In his presidential campaign, Senator John McCain proposed building dozens of new nuclear power plants to handle

the country's growing demand for electricity, and he's not alone.[33] Nuclear power doesn't contribute to global warming, which is a big plus, and it could be an affordable, reliable source of electricity, if only it didn't have such a long, complicated, and expensive birth. Some of the gestation time is used to locate a suitable site and complete the research and contingency planning needed to file an application with the Nuclear Regulatory Commission. Then, of course, it takes time to build a large, complicated structure such as a nuclear reactor, and honestly, you wouldn't want them rushing the job. But just getting an approval from the NRC can take about four years, adding even more time to an already long and costly process.[34] Many advocates of more nuclear power want to streamline and simplify the process. Experts at the Heritage Foundation describe a fast-track approach as one that would cut the approval time in half for reactors that use an NRC-certified design and are located on existing nuclear plants operated by experienced nuclear operators.[35] It's a little like the registered traveler program at airport security—if all your papers are in order, you've been checked out dozens of times before, and you're doing what you always do, there's really no need to start from scratch again.

If you like nuclear power, this seems like a sensible and helpful approach. If you don't, it seems reckless and irresponsible. Opponents say that with more nuclear plants operating, the chances of a serious accident or a security lapse are even greater. Plus, the United States still hasn't agreed on a long-term strategy for storing nuclear waste, and with more plants, of course, there would be even more of that waste to store.

Require that a set amount of America's energy come from renewable sources. More than half of the states have something called a "renewables portfolio standard," affectionately known as the RPS. (Who names these things anyway?) An RPS requires the state's utilities to get a certain amount of energy from renewable sources by a certain date. For example, Rhode Island sets the target at 16 percent of the state's energy by 2019; New Mexico is similar, with a target of 20 percent by 2020. Maine (40 percent by 2017) and California (20 percent by 2010) are more ambitious.[36] The exact rules vary (and some states' standards are nonbinding), but a growing number of experts say it's time to do the same thing nationally. President Obama has talked about having 25 percent of the country's energy from renewable sources by 2025.[37] What's happened around the country suggests some of the pros and cons to the idea. A study by the Lawrence Berkeley National Laboratory showed that an RPS leads to greater use of renewables, especially wind.[38] In most states, electricity rates haven't gone up shockingly, at least so far.[39] Some states have had trouble enforcing their RPS, and some have had transmission problems, for example, getting the energy from the wind farm to the city.[40]

There are lots of disputes about the cost to consumers, and it would differ from place to place.[41] The Bush White House opposed a national RPS in 2007, saying it would be unfair to states without good alternative energy sources.[42] There's always the mandate-versus-incentive debate. But the biggest debate now seems to be what's counted and what's not. RPS targets can generally be met with wind, solar, and geothermal power, but some argue that nuclear power and clean coal technology should be counted since they too reduce greenhouse gas emissions.[43] Experts from MIT who studied the prospects for nuclear and clean coal technologies argue that a national RPS should be a level playing field—one that includes any form of low-carbon energy, including these two. If you've been paying attention at all, we're sure you can fill in the arguments back and forth from there.

Use alternative fuels in cars. Remember our three goals for energy—reducing imports, reducing greenhouse gas emissions, and having reliable, predictable supplies. Unfortunately, using diesel and gasoline in cars falls short on all three. We have to import oil; it's a fossil fuel that contributes to global warming, and lately the price has been swinging around like crazy. You can run cars on alternative fuels such as liquefied natural gas (LNG) and biofuels, so you'll hear people say that if we did that more, the United States would be much better off. You may have seen commercials where T. Boone Pickens promotes LNG, Texas drawl and all. Pickens's website for the Pickens Plan points out that emissions from LNG are 23 percent lower than from diesel and 30 percent lower than from gas.[44] Plus the United States has a better supply of natural gas than oil. Then there's biofuels such as ethanol, which are renewable since they are made from crops that include corn and sugarcane. While using them releases some carbon dioxide, growing the plants to make them helps eat it up, so it's more of a wash.[45] So both LNG and biofuels would be better for the environment than gasoline. They would help us reduce our reliance on imported oil. Best of all, the technology exists right now.

So what's the downside? If you've looked through Chapter 12, you already know a few of them. There are millions of cars in the United States but relatively few of them run on alternative fuels. Flex-fuel cars are more expensive. Then, even if you have the car, there's the filling station problem. Right now, fewer than 1,400 stations nationwide offer E85 ethanol (the most environmentally friendly version), and most are in the Midwest.[46] That's not good if you're driving from Phoenix to Albuquerque. Some experts also worry that relying on biofuels will raise food costs.

Another question, practically speaking, is whether corn ethanol (which is what we mainly produce here) can ever be much of a player. It takes a lot of corn to make the stuff. Right now, we use about a fifth of the nation's corn crop to produce a tiny 5 percent of our liquid fuel needs.[47] Then there's the question of priorities. Do we put our money and effort into encouraging biofuels or LNG, or would it be better to focus on the electric cars now in development? Or maybe, as some experts suggest, we should really go for all three—anything to get us off the gas.

Get going on the electric cars. When news reports emerged in 2009 that automaker Tesla hoped to ramp up its production of battery-powered cars if it could get a start-up loan from the Department of Energy, many people took it as a sign that the time for electric cars has finally arrived.[48] GE is bringing out its Chevy Volt plug-in hybrid for 2010; word has it that Nissan is planning to bring a battery-powered car to the market; Tesla seems poised to expand.[49] Getting more battery-powered cars on the market could help reduce the country's dependence on foreign oil, and for many environmentalists, these are the cars that can really save the planet. Unlike LNG-powered cars, they emit practically no greenhouse gases. Unlike those that run on biofuels, they could be powered on anything we can use to make electricity; we wouldn't have to grow crops or drill oil to make our cars go.

That's all well and good, of course, but many people question whether we'll really have a fleet of battery-powered cars anytime soon. The $100,000 price tag on the spiffy, two-seat Tesla Roadster seems more suited to a few Hollywood stars and Wall Street financiers (the ones who aren't bankrupt, that is) than your average Joe; even the more moderately priced Volt will still probably be more expensive than a conventional car. Plus there's a bit of a chicken-

and-egg problem here. People are going to be leery of buying battery-powered cars unless they're affordable and unless you don't have to worry about where and how you're going to recharge the battery, yet the cars will stay expensive and fairly exotic unless increasing numbers of Americans decide to take the plunge. But Americans do have a way to taking to new technology in surprising and sometimes unpredictable ways. Flat-screen TVs used to be expensive and confined to the fairly well-off, but now they dominate the market.

Rebuild America to be more energy-efficient. Since buildings account for about 40 percent of the country's energy use and carbon dioxide emissions, there's a growing group of people who think it's time for some major home improvement.[50] Some of the focus is on changing building codes to make new structures more energy-efficient. The Alliance to Save Energy, a nonprofit group backed by a whole slew of major U.S. corporations including Whirlpool, Home Depot, BP, and Honeywell, says that improved building codes could save consumers $25 billion on their energy bills over the next two decades and cut greenhouse gas emissions as much as taking 28 million cars off the road.[51] There are also ideas for prodding home owners and landlords to replace drafty windows and switch out older furnaces, air conditioners, and refrigerators for newer, more energy-efficient models. So far, the prodding has mainly been in the form of tax incentives and consumer education, but President Obama included almost $8 billion in the American Recovery and Reinvestment Act for the federal Weatherization Assistance Program and the State Energy Program. The administration argues that the spending will "put 87,000 Americans to work and save families hundreds of dollars per year on their energy bills."[52]

What's not to like about this approach? Well, it does

cost money, and those costs show up in increased housing prices and rents, more government spending, and reduced tax revenue (from giving tax breaks for improvements). Opponents often worry about adding more regulation and red tape to the already cumbersome process of building and renovation. Some aren't eager to have the federal government passing energy rules and setting efficiency standards, although appliance and building parts manufacturers typically prefer a single national standard to a slew of different state ones. One problem with putting too many eggs in this particular basket is that while it helps reduce greenhouse gas emissions, it does very little for reducing the country's reliance on foreign oil. Oil is mainly used in transportation, not in housing and commercial buildings, so these ideas don't really help out much on that score.

Organize a "Manhattan Project" on energy. In the darkest days of World War II, physicist J. Robert Oppenheimer headed up a team of American and British scientists who developed the atomic bomb. The Manhattan Project was a who's who of twentieth-century physics: Luis Alvarez, Hans Bethe, Enrico Fermi, Richard Feynman, Edward Teller, Robert Wilson, and of course Oppenheimer himself all worked on the effort. Most historians say the Manhattan Project was pivotal to winning World War II and that it laid the groundwork for U.S. scientific and military preeminence for the rest of the century.[53]

Is this what we need to solve the country's energy problems? A team of sci-tech all-stars? It's an intriguing idea. One model might be MIT's Energy Initiative, which brings together natural scientists, engineers, and social scientists from across MIT to "envision the best energy policies for the future."[54] They work on nuclear fission, biofuels, coal sequestration, solar and geothermal energy, and more. MIT has already authored useful assessments of the future of coal and nuclear power by bringing experts from different departments together to make a joint appraisal. A Manhattan Project would up the ante by pulling in experts from across the country (not just MIT, as impressive as its experts are) and injecting substantial government money to fund the work.

Despite the success of the Manhattan Project (and the success of the Apollo project which pulled sci-tech talent together to put a man on the moon in the 1960s), there are questions. Those two initiatives were focused on very specific engineering problems: making a bomb or putting a man on the moon. There's not much agreement on what those specific problems would be at this point.[55] Do we put these folks on better electric batteries? Lighter hydrogen fuel cells? Biofuels from nonfood crops, such as switchgrass? Until we define the problem, we can't craft the solution. Such a project would also be costly. The Apollo space program cost about $100 billion in current dollars.[56]

Some critics also charge that the federal government does a lousy job of setting the agenda for research and scientific development, and that this is something much better done in the private sector from the ground up. Many also fear that the federal funding would inevitably be influenced by politics, steered toward those with powerful Washington friends or perhaps narrowed to avoid upsetting key political constituencies such as voters in states where coal or ethanol are economically important.

Rebuild the electricity grid. We're not supposed to push any of the ideas here, but this one is a complete no-brainer. Experts on nearly every side of the energy issue agree that our current grid can't handle the changes we're going to need to make, such as moving to renewable electricity or electric cars. Not to mention that the country's power grid is already getting long in the tooth; it may not even be able to keep up with demand as it is. If we don't improve it, we really will be asking, "Who turned out the lights?" (See "Grid and Bear It" on pages 188–192.)

What happened in California in 2006 is a scary preview of the risks. It was an exceptionally hot summer. People understandably left air conditioners on day and night. Demand for power reached record levels, and transformers started failing. More than a million people lost power, and the power outage was blamed for more than 100 deaths.[57] This is why we have to get our act together on this.

What would be needed? We need a grid that can handle the peaks and valleys of wind and solar power. We need a system that can send power from the East Coast to the West Coast after the New Yorkers, Washingtonians, and Floridians have turned off their computers and TVs and gone to sleep. We need a grid that can predict problems before people start calling up and yelling at their local utility because the electricity is out.[58] We need to replace existing copper wires with high-tech models that can carry three to five times as much current. There's not really much debate that we need to do this, and the major hurdle is not going to surprise you: It's likely to cost between $800 billion and $900 billion.[59]

The main issue is who pays and how. In the past, utilities paid for their own transmission lines much the way cities and states take care of their own roads. But we're talking about the equivalent of a major interstate highway system here, so most experts assume that there has to be some federal role. The 2009 economic recovery act included $4.5 billion in matching funds for rebuilding the electric grid.[60] However, some experts caution that the financial hurdles may not be as daunting as the process of overcoming local objections and getting approvals for building new power lines in states and communities nationwide. Could NIMBY defeat the grid? (If you don't know NIMBY, it's on page 11.)[61]

Have the federal government go green. Senator John McCain and President Obama were on opposite sides of the fence on a lot of issues during the 2008 campaign, but the idea that the federal government should use the cleanest energy in the most efficient way possible was one idea they both supported. Senator John McCain pointed out that the U.S. government buys 60,000 vehicles every year, not including the ones used by the military,[62] and that it's "the single largest consumer of electricity in the world."[63] The U.S. Public Buildings Service, which arranges for office space for federal employees, is "the largest public real estate organization in the country," providing "workspace and workplace solutions to more than 100 federal agencies representing over a million federal civilian workers in 2,000 American communities."[64] You get the picture: The U.S. government is really, really big. So the idea is that it should do everything possible to be more energy efficient and reduce emissions. In Senator McCain's words, this would "save taxpayers billions of dollars in energy costs and move the market in the direction of green technology."

The specifics are pretty clear. New federal buildings would have to be built to state-of-the-art energy efficiency standards. The government would buy electricity generated from less environmentally damaging sources like wind, solar, and nuclear. Instead of buying vehicles with traditional combustion engines, the government would order up hybrids, flex-fuel vehicles, cars that run on liquefied natural gas, and if they're available, battery-powered ones too. Given the government's size, this would certainly open the door for entrepreneurs trying to find markets for green technology. You could argue that since the country really needs

to get away from fossil fuels and imported oil, you shouldn't
use taxpayer dollars to feed bad old habits. President
Obama has stated his goal explicitly. He wants to increase
energy efficiency in all new federal buildings by 40 percent
and get a 25 percent increase in current buildings within
five years.[65]

So what's the problem? Do we really have to tell you?
There just isn't any free lunch. Greener vehicles generally
cost more than conventional ones, just as electricity from
renewables costs more than fossil fuels. Refurbishing federal property is just like refurbishing your own home. You
have to pay someone to come in and replace the furnace
and the windows and the air conditioner and do all that
caulking. A far more daunting problem is that, at least in
the vehicle category, this idea hasn't really worked out so
well so far. As of 2008, the government had a fleet of more
than 100, 000 vehicles that can run on alternative fuels. Yet
according to a study by the *Washington Post*, the vehicles
are running on regular gas over 90 percent of the time.[66]

In some respects, the feds face the same problem regular drivers face. Having a flex-fuel car is one thing; having
the alternative fuel nearby to fill it up is another. Ethanol
fuels and liquid natural gas just aren't as widely available
as regular gasoline. There's some evidence that the government could do better with better planning. According to
the *Post*, thousands of the federal vehicles were dispatched
to states that don't have any service stations offering the
alternative fuels.[67] For example, the U.S. Navy, which also
follows the guidelines, operates nearly 700 flex-fuel vehicles
in Hawaii, but as of 2008, only traditional fuels were available on the islands.[68] Like many of the other options outlined here, there's a big difference between having a good
idea and making it work in the real world.

Invest in railroads, mass transit, and alternative jet fuels—and get the kinks worked out of teleconferencing. The energy efficiency advantages of trains, mass transit, and planes are obvious—you can move dozens of people from place to place on the same amount of fuel, and if that fuel is less harmful to the environment, even better. Many cities' mass transit systems use alternative fuels already; rail travel is very fuel-efficient; and the airlines are experimenting with non-petroleum-based fuels as well. It's actually in their self-interest, since they seem to peer into the canyons of bankruptcy every time oil prices spike. And the teleconferencing? People wouldn't have to travel—it's the dream of tired business travelers everywhere.

In theory, most Americans like the idea of having more trains and mass transit.* More than eight in ten say the country should invest more in railroads; seven in ten say more tax dollars should be spent on public transportation such as bus and rail systems.[69] That's surprising, since as a group, we spend way more time in cars than on buses, trains, and planes combined.[70]

Advocates say more Americans would use mass transit and trains if they were more widely available and made sense economically. Ridership jumped in 2008 when gas prices spiked, showing that people are open to the option. Plus, mass transit is a winner when it comes to reducing emissions and use of oil. Even now, public transportation saves some 37 million metric tons of carbon emissions

---

* Maybe we think of trains as places to fall in love—as in Gene Wilder and Jill Clayburgh in *Silver Streak* and Bing Crosby and Rosemary Clooney in *White Christmas*, not to mention Anna Karenina and Count Vronsky, although Anna and the count actually met at the station and the whole thing came to a very bad end.

annually, according to the American Public Transportation Association—the equivalent, they say, of having every "household in New York City, Washington, D.C., Atlanta, Denver, and Los Angeles combined" completely stop using electricity. Another favorite stat of transit advocates: an average transit rider uses about half of the oil of someone in a car.[71] Suppose, advocates say, you could double the number of people riding the rails and getting on public buses.

The drawbacks? You know this one without us even saying it. All together now—the money! Mass transit systems need tax subsidies and require very large initial public investments. According to the Department of Transportation, "with no subsidies, Amtrak would quickly enter bankruptcy," although the officials also noted that the system could be more cost-effectively reorganized "with service on routes where there is real ridership demand and support from local governments—such as the Northeast Corridor."[72] For many Americans, local mass transit barely exists, and if it does, it just doesn't fit in with their lives. The 2009 economic recovery act includes more than $8 billion for public transit, partly because it meshes with its strategy of launching public works projects that create jobs. One consideration that hasn't gotten much discussion is that more communities may want more public transportation as the population ages. The little old lady from Pasadena may have loved zipping all over town in a fast car, but many older Americans would rather take the bus once they no longer feel comfortable driving.

SO MANY CHOICES, SO LITTLE TIME

At the beginning of this chapter, we cautioned that we'll have to pursue a number of different strategies to make progress. It's also true that we can't possibly afford to do them all. The trick here may be to do what we've haven't done very well in the past: Pick some options that seem sensible, allocate the money they need, and then really try to make a go of them. After all, a half-baked plan isn't all that much better than no plan at all.

This Little Piggy Went to Market; This Little Piggy Passed Laws

Here's a question that underlies much of the debate on energy: Is it better to rely on government action or the marketplace? We've avoided giving unconditional answers to most energy and environmental questions because you deserve the chance to make up your own mind, but in this case, the answer is obvious. It's both. The U.S. economy relies on both government and the marketplace in virtually every other area. Why would energy be any different?

There are good reasons to try to find the golden mean on this question and avoid going too far either way. Absolute devotion to the marketplace gave us the Irish potato famine, when government officials decided that providing food aid might undercut farmers' incentive to work. Absolute faith in government—well, you don't have to go much further than the Soviets and their five-year plans and Castro's Cuba to understand what happens when principles of competition, supply and demand, and risk and reward are thrown completely out the window. So despite some huffing and puffing on the op-ed pages, the United States is not going to do it completely one way or the other.

On the other hand, there are important tensions between a mainly regulatory approach and one that relies mainly on the marketplace, so it's not a bad idea to review some major arguments on both sides. Here's a cheat sheet for you in case your Econ 101 has gotten rusty.

THOSE WHO PUSH FOR MORE REGULATION AND GOVERNMENT ACTION OFTEN MAKE ARGUMENTS LIKE THESE:

1. The market works very well when costs and benefits are immediate and clear, but it doesn't work as well when the costs are mainly in the future (as in global warming) or when the benefits are far down the road (as in investing in the science and technology R & D).
2. The market works very well when the people conducting the transactions pay the costs and gain the benefits personally (because people actually are pretty smart). It doesn't work as well when larger social benefits and costs are in play. For most of us, as private individuals, finding the cheapest form of energy and the most attractive and affordable products is our top priority. But for American society as a whole, reducing imports of foreign oil, protecting nature and wildlife, and working to combat global warming are competing concerns.
3. Government should step in when the private sector and marketplace fail to serve the public good, and the United States has been far too slow to respond to the dangers of global warming and overreliance on fossil fuels.

THOSE WHO WANT LESS GOVERNMENT REGULATION AND MORE RELIANCE ON THE MARKET OFTEN MAKE ARGUMENTS LIKE THESE:

1. Government regulation is bureaucratic and wasteful, and it adds to the cost and difficulty of change. On energy, for instance, Congress at one point asked the Government Accountability Office to make a list of all the federal programs

working on energy. The count: eighteen federal agencies, more than 150 programs, and eleven tax preferences that focus on everything from how much energy we have to helping low-income consumers with energy costs.[73]

2. Legislation and government regulation are usually shaped by politics and lobbying—not the best thinking on how to solve a problem. Competition in the marketplace is a much better strategy for surfacing the best ideas and solutions.

3. In the end, new forms of energy and more energy-efficient products have to be commercially viable. We live in a market economy, and taxpayers shouldn't be put in the position of subsidizing ideas that can't stand on their own.

Both sides can point to persuasive examples to make their point. Advocates of government action often bring the legacy of the environmental movement to the table. For many Americans, laws such as the Wilderness Act, the Clean Air Act, and the Endangered Species Act are fundamental government remedies that are needed to protect the interests of the American public overall.

Yet, what some see as progress, others see as a costly and counterproductive detour. Dan Kish of the Institute for Energy Research believes that government regulation over the past forty years has deepened the country's energy problems and that it's "time for American citizens to take back the energy their government will not allow them to use."[74] Kish and many others argue that "it is time to use our own supplies, and America has no shortage. We simply lack the political will to push government aside and put Americans to work producing them."[75]

A MEETING OF THE MINDS?

Even though the two sides seem poles apart, some key advances rest on both government and market action. *Washington Post* columnist Michael Gerson argues that it was the interweaving of government regulation and private sector ingenuity that helped the country reduce air pollution.[76] "Smog was reduced mainly by innovation, not austerity," Gerson writes, pointing out that the catalytic converter and reformulated gasoline made the advances possible. "But liberals are correct about something else," he adds. "This technological progress would not have taken place as a result of the free market alone.... Federal and state regulations on auto emissions and air quality created an environment in which the invention of new technologies was economically necessary."[77]

Government regulation and market innovation were a potent combination in the country's fight against air pollution. Neither would have been as effective on their own. Photo by Warren Gretz. Courtesy of the U. S. Department of Energy, National Renewable Energy Laboratory

Some predict that an energy and environmental middle ground is emerging. T. Boone Pickens is a tough-minded, tough-talking entrepreneur straight from central casting, and his conservative political background is either legendary or infamous, depending on how you look at the world. Yet he is forceful in saying that the private sector alone cannot address the country's problems with imported oil. "Waiting for the free market to act can be disastrous," Pickens has said.[78]

In September 2007, sixteen major environmental groups including the Sierra Club, Wilderness Society, National Audubon Society, and Greenpeace signed a letter to the U.S. Senate calling for action on global warming. Nothing unusual in that—these groups have long sought government solutions to protect the environmental interests they care about. But the letter also calls for "market flexibility for cost-effective solutions," and national policies that "encourage efficiency, innovation, competition and fairness."[79]

Maybe it's all window dressing. It wouldn't be the first time people with different agendas adopted the best phrases of their opponents to further their own cause. But then again, maybe there is beginning to be some meeting of the minds.

President Obama's Promises

President Obama probably had pretty good report cards all through school, but how's he doing on energy? Courtesy of the U.S. Department of Energy

Like nearly all candidates for office, President Obama made some promises before he was elected about what he would do to address the country's energy and environmental problems. In the 2008 presidential campaign, energy and the environment weren't top issues, but that doesn't mean that the winner shouldn't be graded on how well he does what he said he would do. We're not suggesting picking on the president alone. It's worth taking a look at what your senators and representatives said, along with what your governor and local elected officials promised too. We need to hold all our elected officials accountable for progress in this area. But one of the perks of being elected president is that people remember to watch what you do pretty carefully, so here's your chance to grade the prez on how he's doing.

We're going to use the president's own words as a starting point, and we've included some reminders about what Obama has done so far (or at least before this book went to the printer). You can use the standard A-to-F system we've all come to know and love. In this instance, we're suggesting:

- Use A or B if you like what the president said and think he's making progress
- Use C if you think the idea is fine but not much is happening
- Use D or F if you think the president is on the wrong track

What President Obama Said	Since He Said It	Your Grade
"To finally spark the creation of a clean-energy economy, we will double the production of alternative energy in the next three years."[80]	The president's 2010 budget submitted to Congress calls for $150 billion in spending in alternatives over ten years.	
"We will modernize more than 75 percent of federal buildings and improve the energy efficiency of 2 million American homes, saving consumers and taxpayers billions on our energy bills."[81]	The 2009 Recovery Act includes a weatherization program aimed at "modernizing 75 percent of federal building space and more than one million homes."[82]	
"And I'll invest 150 billion dollars over the next decade in affordable, renewable sources of energy—wind power and solar power and the next generation of biofuels."[83]	The president's 2010 budget submitted to Congress calls for $150 billion in spending in alternatives over ten years.	
"We can't be afraid to stand up to the oil and auto industry when the future of our economy is at stake. When we let these companies off the hook; when we tell them they don't have to build fuel-efficient cars or transition to renewable fuels, it may boost their short-term profits, but it is killing their long-term chances for survival and threatening too many American jobs."[84]	The president's 2010 budget proposes repealing a number of tax benefits for oil exploration and recovery. It also includes a tax on oil and gas from the Gulf of Mexico.	
"I'll help our auto companies retool, so that the fuel-efficient cars of the future are built right here in America. I'll make it easier for the American people to afford these new cars."[85]	The U.S. government now owns General Motors, although to be fair that wasn't part of anyone's plan during the election campaign. Congress also passed a "cash for clunkers" program, offering up to $4,500 for those who trade in older cars that get less than 18 mpg for new fuel-efficient models.	

What President Obama Said	Since He Said It	Your Grade
"As president, I will set a hard cap on all carbon emissions at a level that scientists say is necessary to curb global warming—an 80 percent reduction by 2050. To ensure this isn't just talk, I will also commit to interim targets toward this goal in 2020, 2030, and 2040."[86]	The president called for a cap-and-trade law in February 2009. The House passed a version of this a few months later.	
"That's why we must invest in clean coal technologies that we can use at home and share with the world. Until those technologies are available, I will rely on the carbon cap and whatever tools are necessary to stop new dirty coal plants from being built in America—including a ban on new traditional coal facilities."[87]	The stimulus package and 2010 budget both contain money for clean coal development.	
"We will also explore safer ways to use nuclear power, which right now accounts for more than 70 percent of our non-carbon-generated electricity. We should accelerate research into technologies that will allow for the safe, secure treatment of nuclear waste. As president, I'll continue the work I began in the Senate to ensure that all nuclear material is stored, secured, and accounted for—both at home and around the world."[88]	The administration cut money for nuclear storage at Yucca Mountain, promising to develop a better plan.	

CHAPTER 16

THE REALITY SHOW

Even if you're on the right track, you'll get run over if
you just sit there.

—*Will Rogers*

Don't worry. This "reality show" is not about the trials
and tribulations of Kim and Kourtney Kardashian or
all those people who seem so astonishingly eager to work
for Donald Trump or P. Diddy. (No, we don't understand
it, either.) This reality show is about the facts we have to
face in order to solve the country's energy problems. First,
we need to get ourselves in a problem-solving frame of
mind, which might be helpful in addressing some of the
country's other problems as well. We need to be willing to
listen and learn, and we have to be ready to compromise
and adjust. We need to let go of the finger pointing that's
been so characteristic of the energy issue in the past. We
have to identify and support elected officials who care
enough about the country's future to make some less than
totally fun decisions now. In other words, we need to do
the exact opposite of what reality show contestants usually do. Temper tantrums and hysterics, unlike energy,
are cheap, plentiful, and not very helpful.

To change our ways, here are some of the realities we need to keep in mind.

Reality #1: Changing Lightbulbs Is Not Enough

It's a fine thing to do. If you go through your house changing all the incandescent bulbs to the more efficient compact fluorescents, you'll save some energy. If you can get your entire extended family and all your friends and neighbors to do it too, so much the better. But it's not enough—not even if we outlaw the old-style bulbs and send armed search teams throughout the land confiscating them from the obstinate. To ensure that this country has safe, reliable, and affordable energy into the future, we need to make larger decisions and investments, and we need to make changes that add up to big numbers.

Reality #2: We'll Have to Live for Years with Decisions We Make Now

Continuing to make the same decisions we've made in the past will put the United States into an even deeper energy and environmental hole. You've heard us say it takes a long time to get anything done in the energy world, but it also takes a long time to undo decisions. A power plant typically operates for three or four decades. Americans will be living in the buildings we build now for the next 40, 50, or 150 years. Some of the greenhouse gases we're pumping into the atmosphere now will be there for centuries.[1] The decisions we make now can put us in good stead for the future—or they can make our problems worse.

Reality #3: One Piece of Legislation Won't Do It

Congress has passed several important energy acts of late, and chances are it will pass some more over the next few years. But steps such as these are just the beginning. Transforming the way the United States gets and uses energy depends on reforms and adaptation in nearly every direction—changes in federal, state, and local laws; zoning modifications; marketplace innovations; shifts in product availability; alterations in Americans' habits and expectations. This is a huge social and economic shift, and we won't be done when the president signs the next energy bill that comes down the pike. We'll have just taken another step along the way.

Reality #4: Fossil Fuels—Can't Live with Them, Can't Live Without Them (at Least Not Yet)

Americans need to start thinking differently about fossil fuels, especially oil. At the same time, fossil fuels supply 80 percent of the country's energy now, and that's not something we can change overnight; it's just not possible.[2] In effect, the United States needs to start moving away from fossil fuels at the same time that we're still depending on them. It's a little like a story that's been making the rounds, reportedly one John F. Kennedy used to tell. A man asks his gardener to plant a tree, and the gardener complains that it will be fifty years before the tree begins to look like anything. "Well, you better plant it this morning," the man says.[3] That's about where we are with fossil fuels. We need to start now, so that we'll be in a different situation fifty years from now.

Reality #5: Cheap Energy Is a Blessing and a Curse

There's no question that low energy prices over the last two decades have been a boon to the U.S. economy, and the oil price spike of 2008 played its part in starting the 2009 global financial crisis. Plus, cheap energy helps people around the world get out of poverty. (Remember, out of the nearly 6.8 billion people on this planet, fully one-quarter still don't have electricity.) But as long as the fossil fuels we use now are much cheaper than the alternatives, we're going to stall on switching over. It may not be true forever, but right now, clean energy isn't cheap, and cheap energy isn't clean. Unless we find ways to change that reality, we (and the whole world) may procrastinate until it's too late.

Reality #6: A Prototype Is Not a Plan

Sometimes a problem doesn't get solved because no one has any good ideas about what to do. But in energy, the complete opposite is true. There are budding Thomas Edisons popping up everywhere telling us we can get energy from the ocean tides or the Earth's core. Some clever folks are generating energy from garbage and leftover cooking oil, from algae or dirty socks—well, that last one's a fib (at least as far as we know). Someone sets up the experiment, and lo and behold, it works. But showing that something can work doesn't mean that it will work out in real life. Even ideas that are way beyond the experimental stage have to leap a lot of hurdles to get out of the starting gate. They need to be competitively priced. They need to be convenient. They need to enter mainstream, everyday American life, and that just doesn't happen automatically. "Eureka" moments are nice, but making things practical and affordable is going to be just as important.

Reality #7: We Need to Help People Who Lose Out When Energy Habits Change

Election pundits love to divide the country into red and blue states based on political preferences, but some observers are starting to divide the country along "green" and "brown" lines as well.[4] In so-called green states such as California and those on the East Coast, there's greater interest in alternatives, renewables, and refashioning the energy mix. In so-called brown states including Ohio, West Virginia, Michigan, and others, there's more reliance on fossil fuels, greater ties to coal or oil production, and steeper obstacles to moving swiftly to alternatives. These regions face a tougher challenge, and the trade-offs are much sharper for them.[5] So this leads us to our last reality, which is that we Americans may need to think about how to help areas of the country where energy solutions have more serious downsides and individuals who are harder hit when change comes their way. The United States didn't do such a great job along these lines in the shift from a manufacturing economy to an information one. We let some of our great industrial cities decline, and some parts of the country have still not recovered from factory closings and the loss of good-paying jobs. We need to do better this time around.

THE LAST PICTURE SHOW

If you're an *NCIS* fan, you know that Special Agent DiNozzo can come up with a movie analogy for nearly every case the team handles. We've relied on movie analogies quite a bit too, so here's our last batch. As far as we're concerned, there are three movies that foreshadow the country's energy choices.

For some people, the *Mad Max* trilogy is probably the first energy movie to come to mind (We nominate *The Road Warrior* as the best of the three.) Sometime in the future, oil is in very short supply (for reasons that are never fully explained), and Mel Gibson wanders through the Australian outback, where society is coming apart at the seams in the struggle for fuel. In the first movie, the police and hospitals are just barely functioning. By the second and third films, small bands of people barricade themselves around refineries, holding off ruthless biker gangs dressed like 1980s hair bands, only far more vicious.

This is the worst-case scenario—a lurid, pop culture version of the "we're going to run out of oil" school of thought. Whether you're for more domestic drilling or worried about peak oil, *Mad Max* sums up the fear that running out of energy means running out of civilization.

For others, the movie to worry about is 1973's *Soylent Green*, starring Charlton Heston and Edward G. Robinson in his last film role. Here, the Earth has suffered a staggering environmental disaster. New Yorkers are starving in a polluted, corrupt, broiling hot city. A few rich people live well enough in fortresslike apartments, but most people survive by eating "soylent green," a substance that . . . well, we can't tell you what it is without spoiling the plot. Suffice to say that nutritional labeling rules must have been *dramatically* relaxed.

Heston and Robinson are cops, and at one point, Robinson, who is older, talks about how the stores used to be full of food, and you could pick apples off trees. "I know, Sol, you've told me a hundred times before," Heston says. "People were better, the world was better. . . ." Robinson shakes his head. "People were always rotten," he answers, "but the world was beautiful."[6]

Soylent Green imagines what the human and social costs of environmental disaster might be. For many Americans, this movie sums up the fear that basic human decency itself wouldn't survive the strain.

Both *Mad Max* and *Soylent Green* are nightmarish visions of what might happen, but they don't have to be self-fulfilling prophecies. The United States has shown time and time again that this is a country that can change when it's forced to and when it's good and ready. And we've already made some significant progress in the past. In 1973, when *Soylent Green* was made, the country's city dwellers subsisted under a grayish haze while cars running on leaded gas without catalytic converters pumped pollution into the air. At that time, many communities still released minimally treated sewage into rivers and the ocean. Since then, we've done a fair bit. Both the air and the water are cleaner, because government, business, environmental groups, and the public made them cleaner. For all the pain and angst involved, the gasoline price spikes of 2008 showed again what people can do when they see a reason to do it. Americans cut their driving, switched to hybrids, and found all kinds of new energy alternatives. People get creative when they have to.

That's why we keep coming back to a different kind of disaster movie: *Apollo 13*. Based on the actual events in 1970, it tells the story of the Apollo 13 moon mission, which had to be aborted when an oxygen tank exploded in space. This is a movie about problem solving, and while the astronauts are gallant, the real heroes are the NASA engineers on the ground—classic sixties engineers with buzz cuts, short-sleeved white shirts, pocket protectors, and horn-rimmed glasses.

In one sequence, Gary Sinise, playing one of the astronauts left behind, sits in the Apollo simulator trying to figure out a way to bring the spacecraft home on the paltry amount of electricity left in the batteries, barely enough to run a coffeepot. It's a frustrating, painstaking, and tiring effort. He tries dozens of things dozens of times, but he eventually comes up with a plan that ekes out enough voltage to get the ship home. Directing the operation is mission control chief Gene Krantz, played by Ed Harris, who won't let the engineers and scientists fall into recriminations and rivalry. "Let's work the problem, people! Let's not make things worse by guessing,"[7]

The people who brought the Apollo 13 spacecraft back to Earth faced an unexpected and terrifying emergency, but they didn't allow themselves to become immobilized because they were confused. They didn't get sidetracked battling over ideological points. They put their differences aside. They looked over their choices and did the best they could with what they had. In the end, they brought the crew back to Earth safely.

It's not such a bad model for solving the country's energy problems. Maybe it's time to work the problem, people.

ACKNOWLEDGMENTS

We hear stories about authors who complete their books holed away in a garret or a cabin in the woods, pretty much by themselves. That's simply not the case for us. There are many colleagues and friends who advised and helped us along the way.

The approach we've taken in *Who Turned Out the Lights?* and the judgments we've made about how to help typical Americans grasp the energy issue are grounded in the work of Public Agenda's founder, Daniel Yankelovich. Dan, working with Barbara Lee, took a leading role in shaping Public Agenda's Energy Learning Curve research. This book rests on that foundation. Ruth Wooden at Public Agenda and Deborah Wadsworth on our board have been steadfast and enthusiastic supporters. All of our Public Agenda board members and colleagues have been encouraging and supportive. David White has been especially wonderful, helping us cope with some of our technology challenges. Jon Rochkind also provided some much-needed help at key points.

We would also like to thank Jenny Choi, our indispensable fact checker, and Nancy Cunningham who created our charts and graphs. We have benefited from the intelligence and professionalism of Jud Laghi and Larry Kirshbaum at LJK Management. Matthew Inman at HarperCollins has been a remarkably astute editor and reliable sounding board—good-humored and patient through thick and thin. We are also very grateful to our other HarperCollins colleagues: Carrie Kania, Calvert Morgan, Hollis Heimbouch, Lelia Mander, Teresa Brady, Jennifer Hart, Nicole Reardon, Cathy Serpico, Michael Barrs, and Betty Lew. Working with them has been a wonderful experience.

Who Turned Out the Lights? incorporates the insights and counsel of some wonderfully knowledgeable experts who read earlier drafts and gave us extremely helpful advice. William K. Ott, P.E., shared his vast know-how and experience in energy exploration and development with us. Ted Abernathy, executive director of the Southern Growth Policies Board, and his colleagues, Charity Pennock and Linda Hoke, provided enormously useful guidance and feedback. Any mistakes, however, are completely our own.

In a special category of their own, our families and friends have stood by us and cheered us on. Susan Wolfe Bittle and Josu Gallastegui gave us unstinting love and support. We thank them for everything, especially their belief in us.

VISIT OUR WEBSITE

By this point, you may have heard enough about energy, but we suspect quite a few people want to know more. So visit the *Who Turned Out the Lights* website, at www.whoturned outthelights.org, for our appendix of additional resources. If there are things you think we didn't cover, you're probably right. You can find additional features (or if you prefer, the deleted scenes) covering topics like the Strategic Petroleum Reserve, car-buying habits, clean coal in China, and other energy subjects as well. Plus we'll be commenting on new developments as they occur. This is an issue where the news just never stops.

NOTES

PREFACE: WHERE WE'RE COMING FROM

1 See Public Agenda, *The Energy Learning Curve*, April 2009, www .publicagenda.org/reports/energy.

2 See, for example, Cambridge Energy Research Associates, "New CERA/ World Economic Forum report offers broad-ranging discussion on the future of energy innovation, April 15, 2008, www.cera.com/aspx/cda/ public1/news/pressReleases/pressReleaseDetails.aspx?CID=9434.

3 Environmental Protection Agency, Statement of Lisa P. Jackson, "EPA Finds Greenhouse Gases Pose Threat to Public Health, Welfare," April 17, 2009, http://yosemite.epa.gov/opa/admpress.nsf/0/0EF7DF6758052 95D8525759B00566924.

4 www.imdb.com/Find?select=Quotes&for=the%20easy%20way%20or %20the%20hard%20way.

CHAPTER 1. SIX REASONS THE UNITED STATES NEEDS TO GET ITS ENERGY ACT TOGETHER

1 Energy Information Administration, *Energy Consumption, Expenditures, and Emissions Indicators, 1949–2006*, Table 1.5, www.eia.doe.gov/ emeu/aer/txt/ptb0105.html.

2 "New EIA Energy Outlook Projects Flat Oil Consumption to 2030, Slower Growth in Energy Use and Carbon Dioxide Emissions, and Reduced Import Dependence," U.S. Energy Information Administration press release, December 17, 2008, www.eia.doe.gov/neic/press/press 312.html.

3 International Energy Agency, *World Energy Outlook 2008*, www .worldenergyoutlook.org.

4 Ibid.

5 Energy Information Administration, International Energy Outlook 2009, May, 2009. www.eia.doe.gov/oiaf/ieo/highlights.html.

6 See Randy Udall and Steve Andrews, "Oil Shale May Be Fool's Gold," Energy Bulletin, December 17, 2005, www.energybulletin.net/node/

11779, http://planetforlife.com/pdffiles/manifesto.pdf, and "John Mc-Cain's Energy Follies," *New York Times*, September 7, 2008.

7 EIA, "Frequently Asked Questions—Crude Oil. Question: Do we have enough oil worldwide to meet our future needs?" http://tonto.eia.doe.gov/ask/crudeoil_faqs.asp#oil_needs.

8 EIA, *International Energy Outlook 2008*, Chapter 3, "Natural Gas," www.eia.doe.gov/oiaf/ieo/nat_gas.html.

9 Congressional Budget Office, *The Economic Effects of Recent Increases in Energy Prices*, July 2006.

10 Energy Information Office, *Short Term Energy Outlook*, Price Summary, February 9, 2009, www.eia.doe.gov/steo.

11 See, for example, Jad Mouawad, "Rising Fear of a Future Oil Shock," *New York Times*, March 27, 2009.

12 Energy Information Administration, "Natural Gas Navigator: Natural Gas Prices, 2002–2007," http://tonto.eia.doe.gov/dnav/ng/ng_pri_sum_dcu_nus_a.htm.

13 Energy Information Administration, *Uranium Marketing Annual Report*, www.eia.doe.gov/cneaf/nuclear/umar/summaryfig2.html.

14 Energy Information Administration, Frequently Asked Questions, http://tonto.eia.doe.gov/ask/crudeoil_faqs.asp#foreign_oil.

15 Intergovernmental Panel on Climate Change, *Climate Change 2007, Synthesis Report: Summary for Policymakers.*

16 American Association for the Advancement of Science, *AAAS Board Statement on Climate Change*, December 9, 2006.

17 Congressional Budget Office, *Policy Options for Reducing CO_2 Emissions*, February 2008.

18 Steve Gelsi, "Exxon Mobil Redirects Public-Policy Research Funds," MarketWatch.com, May 27, 2008, www.marketwatch.com/news/story/exxon-mobil-redirects-climate-change/story.aspx?guid=%7BE8735291%2DC763%2D4A82%2D982A%2DB5E38CF2D571%7D&dist=msr_3; ExxonMobil, "Rising to the CO_2 Challenge," ad in *New York Times*, February 12, 2009.

19 Charles Krauthammer, "Confessions of a Global-Warming Agnostic," *National Review*, May 30, 2008, http://article.nationalreview.com/?q=ZGI0MDdiZDQ3MGI1ZGYzNWZkzTcwZWM5YzI2MWI5N2U=.

20 OPEC, "Frequently Asked Questions: Who Are OPEC Member Countries?" www.opec.org/library/FAQs/aboutOPEC/q3.htm.

21 OPEC, "Frequently Asked Questions: Does OPEC Control the Oil Market?" www.opec.org/library/FAQs/aboutOPEC/q13.htm.

22 Jad Mouawad, "Exxon Profit Down 33% as Prices Fall," *New York Times*, January 30, 2009, www.nytimes.com/2009/01/31/business/31oil.html?_r=1&hp.

23 The Polling Report, USA Today/Gallup Poll, May 30-June 1, 2008. "Please tell me whether you think each of the following deserves a great deal of blame, some blame, not much blame, or no blame at all for the country's current energy problems. How about U.S. oil compa-

nies? A great deal, 60 percent; some, 30 percent; not much, 5 percent; none, 4 percent." www.pollingreport.com/energy.htm.

24 Paul Krugman, "Fuels on the Hill," *New York Times,* June 27, 2008.

25 Energy Information Administration, "Impacts of Increased Access to Oil and Natural Gas Resources in the Lower 48 Federal Outer Continental Shelf," 2007, www.eia.doe.gov/oiaf/aeo/otheranalysis/ongr.html.

26 Environmental Protection Agency, "Light-Duty Automotive Technology and Fuel Economy Trends: 1975 through 2007," September 2007, Table 3: "Sales Fractions of MY1975, MY1988 and MY2007, Light-Duty Vehicles by Vehicle Size and Type." www.epa.gov/otaq/fetrends .htm.

27 Energy Information Administration, Crude Oil: Frequently Asked Questions, When Was the Last Refinery Built in the United States? http://tonto.eia.doe.gov/ask/crudeoil_faqs.asp#last_refinery_built

28 EIA Kids Page, "Nuclear Energy (Uranium), Energy from Atoms," www .eia.doe.gov/kids/energyfacts/sources/non-renewable/nuclear.html.

29 International Atomic Energy Agency, "Nuclear Power Worldwide: Status and Outlook," State News Service, September 11, 2008, www.iaea .org/NewsCenter/PressReleases/2008/prn200811.html.

30 EIA Kids Page, "Nuclear Energy (Uranium), Energy from Atoms."

31 See, for example, Global Nuclear Energy Partnership, www.energy .gov/media/GNEP/06-GA50035b.pdf.

32 Based on an analysis by PricewaterhouseCoopers presented at the 2005 Global Energy, Utilities and Mining Conference, November 16–17, 2005, www.pwc.com/extweb/industry.nsf/docid/49f2db1ed1e b0236852571c6005adc63/$file/tom-collins-noc-presentation-for-web-site.pdf. This analysis and others are available at the Energy Information Administration's Web page "Energy-in-Brief: Who Are the Major Players Supplying the World Oil Market?" accessed April 2, 2009, http://tonto.eia.doe.gov/energy_in_brief/world_oil_market.cfm.

33 EIA, www.eia.doe.gov/pub/international/iealf/tableh1co2.xls.

34 World Resources Institute, *Navigating the Numbers: Greenhouse Gas Data and International Climate Policy,* http://pdf.wri.org/navigating_ numbers_chapter6.pdf.

35 EIA, "Frequently Asked Questions—Environment: How Much CO_2 Does the United States Emit? Is It More than Other Countries?" updated August 14, 2008, http://tonto.eia.doe.gov/ask/environment_faqs .asp#greenhouse_gases_definition.

36 EIA, *Emissions. of Greenhouse Gases Report: U.S. Emissions in a Global Perspective,* Report #DOE/EIA-0573, December 3, 2008, www .eia.doe.gov/oiaf/1605/ggrpt/index.html.

37 OPEC, "Frequently Asked Questions: Does OPEC Set Crude Oil Prices?" www.opec.org/library/FAQs/aboutOPEC/q20.htm.

38 www.nymex.com/CL_spec.aspx.

39 OPEC, "Frequently Asked Questions: Does OPEC Set Crude Oil Prices?"

40 CIA, *The World Factbook 2008,* www.cia.gov/library/publications/the-world-factbook/geos/us.html.

41 BP, *Statistical Review of World Energy*, 2009. www.bp.com/liveassets/ bp_internet/globalbp/globalbp_uk_english/reports_and_publications/ statistical_energy_review_2008/STAGING/local_assets/2009_down loads/statistical_review_of_world_energy_full_report_2009.pdf.

42 Ibid.

43 Ibid.

44 See, for example, www.nytimes.com/2007/02/02/science/earth/02cnd-climate.html.

45 National Science and Technology Council, *Scientific Assessment of the Effects of Global Change on the United States, Report of the Committee on Environment and Natural Resources*, May 29, 2008. www.climate science.gov/Library/scientific-assessment.

46 Andrew C. Revkin, "Under Pressure, White House Issues Climate Change Report," *New York Times*, May 30, 2008, www.nytimes.com/ 2008/05/30/washington/30climate.html?_r=1&ref=us&oref=slogin.

47 www.washingtonpost.com/wp-dyn/content/article/2008/01/20/AR200 8012002388_2.html.

48 In 2007, the U.S. gross domestic product (the total size of the economy) grew at a very respectable 2 percent rate, but China's economy grew at a staggering 11.9 percent. International Monetary Fund, *World Economic Outlook*, October 2008, www.imf.org/external/pubs/ft/weo/ 2008/02/index.htm.

49 International Energy Agency, *World Energy Outlook 2007*, "China and India Insights," www.worldenergyoutlook.org/2007.asp.

CHAPTER 2: GROUNDHOG DAY, OR HAVEN'T WE SEEN THIS MOVIE BEFORE?

1 Pew Charitable Trusts, "History of Fuel Economy: One Decade of Innovation, Two Decades of Inaction," www.pewtrusts.org.

2 Ibid.

3 Joseph B. White, "Can Big, Fast Cars Be Eco-Friendly?" *Wall Street Journal*, April 25, 2008, http://online.wsj.com/article/SB120908995550 343947.html?mod=hpp_us_personal_journal.

4 See, for example, John J. Miller, "Unhappy Hour: Daylight Savings Time—A Bad idea," *National Review*, April 1, 2005, http://article.nation alreview.com/?q=N2JhMmQxNjRkNjljZTdmNDM5MzFhNjcwMD c0YjE4ZDE=#more.

5 EIA, *Gasoline and Diesel Fuel Update/History*, http://tonto.eia.doe.gov/ oog/info/gdu/gasdiesel.asp.

6 www.fueleconomy.gov, 2006 Hummer H3, 4WD, www.fueleconomy .gov/Feg/noframes/22514.shtml.

7 *New York Times*, "Chinese Company Buying G.M.'s Hummer Brand," June 2, 2009.

8 Energy Information Agency, Annual Oil Market Chronology, World Nominal Oil Price Chronology: 1970–2007, www.eia.doe.gov/emeu/ cabs/AOMC/Overview.htm

9 "Indepth: Oil, Price Fluctuations: A Timeline," Canadian Broadcasting Corporation, April 18, 2006, www.cbc.ca/news/background/oil/forces.html.

10 U.S. Census Bureau, International Data Base, "Total Midyear Population for the World: 1950–2050," www.census.gov/ipc/www/idb/worldpop.html.

11 Daniel Yergin, chairman IHS Cambridge Energy Research Associates, "The Long Aftershock: Oil and Energy Security After the Price Collapse," Testimony to the Joint Economic Committee of Congress, May 20, 2009, www2.cera.com/dy20090520.pdf.

CHAPTER 3: GIVING THE VOTERS WHAT THEY WANT

1 Opensecrets.org, Influence and Lobbying: Energy/Natural Resources. Based on figures as of May 27, 2009, www.opensecrets.org/lobby/indus.php?year=2008&lname=E&id=.

2 Opensecrets.org, Top Industries Giving to Members of Congress, 2008 Cycle, based on figures as of May 27, 2009, www.opensecrets.org/industries/mems.php?party=A&cycle=2008.

3 Ibid.

4 OpenSecrets.org, Influence and Lobbying, Environment Industry Profile, 2008, www.opensecrets.org/lobby/indusclient.php?lname=Q11&year=2008.

5 OpenSecrets.org, "Influence and Lobbying, Environment: Long-Term Contribution Trends," www.opensecrets.org/industries/indus.php?cycle=2008&ind=Q11. Based on figures as of July, 2009.

6 Lindsay Renick Mayer, "Big Oil, Big Influence," *NOW*, www.pbs.org/now/shows/347/oil-politics.html.

7 Ibid.

8 Public Agenda, "Energy Learning Curve," April 2009, www.publicagenda.org/pages/energy-learning-curve.

9 Ibid.

10 Gallup Organization, "Little Increase in Americans' Global Warming Worries," April 21, 2008. Do you think that global warming will pose a serious threat to you or your way of life in your lifetime? No: 58 percent; Yes: 40 percent. www.gallup.com/poll/106660/Little-Increase-Americans-Global-Warming-Worries.aspx.

11 Ibid.

12 Public Agenda, "Energy Learning Curve."

13 Ibid.

14 Energy Information Administration, "Analysis of Crude Oil Production in the Arctic National Wildlife Refuge," May 2008, www.eia.doe.gov/oiaf/servicerpt/anwr/pdf/sroiaf(2008)03.pdf.

15 BP, *Statistical Review of World Energy*, 2008, "Oil Production," www.bp.com/liveassets/bp_internet/globalbp/globalbp_uk_english/reports_and_publications/statistical_energy_review_2008/STAGING/local_assets/downloads/pdf/oil_table_of_world_oil_production_million_tonnes_2008.pdf.

CHAPTER 4. SEEMED LIKE A GOOD IDEA AT THE TIME, OR HOW THREE FLAWED IDEAS COULD GET US OFF TRACK

1 See the Council on Foreign Relations website at www.foreignpolicy .com/users/login.php?story_id=3262&URL=www.foreignpolicy.com/ story/cms.php?story_id=3262 for Philip J. Deutch's "Think Again: Energy Independence"; the Cato Institute at www.cato.org/pub_display .php?pub_id=4608 for "The Energy Mirage" by Jerry Taylor and Peter Van Doren and Brookings at www.brookings.edu/~/media/Files/rc/ papers/2008/1230_energy_nivola/1230_energy_nivola.pdf for Pietro S. Nivola's "Rethinking 'Energy Independence.'"

2 EIA, Annual Energy Outlook Early Release Overview, Report #DOE/ EIA-0383, accessed March 1, 2009: "As a result, U.S. dependence on imported liquids, measured as a share of U.S. liquids use, is expected to continue declining over the next 25 years, from 58 percent in 2007 to less than 40 percent in 2025, before increasing to 41 percent in 2030." www.eia.doe.gov/oiaf/aeo/overview.html#production.

3 Brookings Institution, "Rethinking 'Energy Independence,'" December 30, 2008, www.brookings.edu/~/media/Files/rc/papers/2008/1230_ energy_nivola/1230_energy_nivola.pdf.

4 Jerry Taylor, "Energy Independence Won't Help Us," *Marketplace*, March 20, 2009, http://marketplace.publicradio.org/display/web/2008/ 03/20/energy_independence.

5 "The Oil Price Needs of OPEC Members," Reuters, November 20, 2008; Luke Harding, Ian Black, and Rory Carroll, "From the Kremlin to Caracas: How the Oil Collapse Changes Everything," *Guardian*, November 21, 2008, www.guardian.co.uk/business/2008/nov/21/oiland gascompanies-globaleconomy.

6 Bill Torpy, "End of the Road Comes Later for Cars," *Atlanta Journal-Constitution*, April 9, 2008.

7 Government Accountability Office, Testimony Before the House Subcommittee on Energy and Environment, Committeee on Science, Technology, Statement of Mark E. Gaffigan, *Advance Energy Technologies, Budget Trends and Challenges for DOE's Energy R & D Program*, www.gao.gov/new.items/d08556t.pdf.

8 Energy Information Administration, "Analysis of Crude Oil Production in the Arctic National Wildlife Refuge," Report #SR-OIAF/ 2008-03, May 2008, p. 2, www.eia.doe.gov/oiaf/servicerpt/anwr/intro duction.html.

9 Department of Energy, "Wind Energy Could Produce 20 Percent of U.S. Electricity by 2030," *DOE Report Analyzes U.S. Wind Resources, Technology Requirements, and Manufacturing, Siting and Transmission Hurdles to Increasing the Use of Clean and Sustainable Wind Power*, May 12, 2008, www.energy.gov/news/6253.htm.

10 Andrew Eder, "TVA Board OKs Completion of Nuke Reactor," Knoxville *News-Sentinel*, August 2, 2007.

11 Toni Johnson, "Challenges for Nuclear Power Expansion," Council on Foreign Relations, August 11, 2008, www.cfr.org/publication/16886.

12 U.S. Nuclear Regulatory Commission, "History of Watts Bar Unit 2 Reactivation," www.nrc.gov/reactors/plant-specific-items/watts-bar/history.html, accessed May 31, 2009.

13 Tennessee Valley Authority, "Watts Bar Nuclear Plant," www.tva.gov/power/nuclear/wattsbar.htm, accessed May 31, 2009.

14 "In the Field: Pilot Project Uses Innovative Process to Capture CO_2 from Flue Gas," *EPRI Journal*, Spring 2008, pp. 4–5, http://mydocs.epri.com/docs/CorporateDocuments/EPRI_Journal/2008-Spring/1016422.pdf.

15 "Growing World Needs Every Form of Energy," April 20, 2008, www.reuters.com.

CHAPTER 5. THE BASICS:
TEN FACTS YOU NEED TO KNOW

1 Energy Information Administration, www.eia.doe.gov/kids/energyfacts/science/formsofenergy.html.

2 Energy Information Administration, www.eia.doe.gov/kids/energyfacts/sources/non-renewable/coal.html#uses;
www.eia.doe.gov/kids/infocardnew.html#ELECTRICITY

3 Energy Information Administration, "Energy in Brief: Electricity," May 9, 2008, http://tonto.eia.doe.gov/energy_in_brief/electricity.cfm.

4 Congressional Research Service, "Energy: Selected Facts and Numbers," May 1, 2008.

5 Energy Information Administration, http://tonto.eia.doe.gov/dnav/pet/hist/rwtcW.htm.

6 Energy Information Administration, www.eia.doe.gov/emeu/steo/pub/fsheets/real_prices.html.

7 U.S. Census Bureau, www.census.gov/popest/archives/1990s/popclockest.txt and www.census.gov/popest/states/NST-ann-est2007.html.

8 Environmental Protection Agency, "Light-Duty Automotive Technology and Fuel Economy Trends: 1975 through 2008," September 2008, www.epa.gov/otaq/fetrends.htm

9 Energy Information Administration, Annual Energy Review, Table 1.5., www.eia.doe.gov/aer/pdf/pages/sec1_13.pdf.

10 U.S. Census Bureau, "Most of Us Still Drive to Work—Alone; Public Transportation Commuters Concentrated in a Handful of Large Cities," press release, June 13, 2007, www.census.gov/Press-Release/www/releases/archives/american_community_survey_acs/010230.html.

11 U.S. Department of Transportation, Bureau of Transportation Statistics, Table 1-11: "Number of U.S. Aircraft, Vehicles, Vessels, and Other Conveyances," www.bts.gov/publications/national_transportation_statistics/html/table_01_11.html.

12 U.S. Department of Transportation, Bureau of Transportation Statistics, Table 1–32: "U.S. Vehicle-Miles," www.bts.gov/publications/na

tional_transportation_statistics/html/table_01_32.html, and Table 4-7: "Domestic Demand for Gasoline by Mode," www.bts.gov/publications/ national_transportation_statistics/html/table_04_07.html.

13 Department of Energy, Buildings Energy Data Book, "Characteristics of a Typical Single-Family Home," Table 2.2.11, www.btscoredatabook .net/TableView.aspx?table=2.2.11.

14 U.S. Energy Information Administration, Regional Energy Profiles, Appliance Reports, U.S. Data Table 2001, accessed May 30, 2009, www .eia.doe.gov/emeu/reps/appli/us_table.html.

15 U.S. Census Bureau, Characteristics of New Housing 2007, Median and Average Square Feet of Floor Area in New One-Family Houses Completed by Location, www.census.gov/const/C25Ann/sftotalme davgsqft.pdf.

16 U.S. Census Bureau, Families and Living Arrangements 2005, www .census.gov/population/www/pop-profile/files/dynamic/FamiliesLA .pdf.

17 Department of Energy, Buildings Energy Data Book, Characteristics of a Typical Single-Family Home, Table 2.2.11., www.btscoredatabook .net/TableView.aspx?table=2.2.11.

18 International Energy Agency, *Key World Energy Statistics 2007*.

19 Energy Information Administration, *Annual Energy Outlook 2008*, March 2008, www.eia.doe.gov/oiaf/aeo/index.html.

20 Public Agenda, "Energy Learning Curve," April 2009, www.public agenda.org/pages/energy-learning-curve.

21 Energy Information Administration, "Energy in Brief: How Dependent Are We on Foreign Oil?," April 23, 2009, http://tonto.eia.doe.gov/ energy_in_brief/foreign_oil_dependence.cfm. Accessed May 16, 2009.

22 Public Agenda, "Energy Learning Curve."

23 U.S. Energy Information Administration, Annual Energy Outlook 2009 Early Release, December 17, 2008, www.eia.doe.gov/oiaf/aeo/ overview.html.

CHAPTER 6. DOUBLE, DOUBLE, OIL, AND TROUBLE

1 It's 122 minutes long; www.imdb.com/title/tt0367882.

2 Energy Information Administration, "Oil/Petroleum," accessed March 29, 2009, www.eia.doe.gov/kids/energyfacts/sources/non-renewable/oil .html#Howformed.

3 Energy Information Administration, "Crude Oil Production," accessed March 29, 2009, http://tonto.eia.doe.gov/dnav/pet/pet_crd_crpdn_adc_ mbblpd_a.htm.

4 International Energy Agency, "Key World Energy Statistics 2008," www.iea.org/textbase/nppdf/free/2008/key_stats_2008.pdf.

5 BP, *Statistical Review of World Energy*, 2008, "Oil Production," www .bp.com/liveassets/bp_internet/globalbp/globalbp_uk_english/reports_ and_publications/statistical_energy_review_2008/STAGING/local_ assets/downloads/pdf/oil_table_of_world_oil_production_million_ tonnes_2008.pdf.

6 U.S. Energy Information Administration, "Crude Oil Production and Crude Oil Well Productivity, 1954–2007," www.eia.doe.gov/aer/txt/ptb0502.html.

7 U.S. Energy Information Administration, "Energy Perspectives: Petroleum Overview and Crude Oil Production," *Annual Energy Review 2006*, www.eia.doe.gov/emeu/aer/ep/ep_frame.html.

8 National Petroleum Council, "Oil Flows and Geographic Choke Points: Facing the Hard Truths about Energy," July 2007.

9 Government Accountability Office, "Energy Markets: Increasing Globalization of Petroleum Products Markets, Tightening Refinery Demand and Supply Balance, and Other Trends Have Implications for U.S. Energy Supply, Prices and Volatility," December 2007.

10 U.S. Energy Information Administration, "Annual Energy Outlook 2007 with Projections to 2030: Oil and Natural Gas," February 2007, www.eia.doe.gov/oiaf/archive/aeo07/gas.html.

11 "Analysis of Crude Oil Production in the Arctic National Wildlife Refuge," Energy Information Administration, May 2008, www.eia.doe.gov/oiaf/servicerpt/anwr/pdf/sroiaf(2008)03.pdf.

12 "This Week in Petroleum," Energy Information Administration, December 17, 2008, http://tonto.eia.doe.gov/oog/info/twip/twiparch/081217/twipprint.html.

13 National Petroleum Council, "Facing the Hard Truths about Energy," July 2007.

14 Ibid.

15 Congressional Research Service, "The Use of Profit by the Five Major Oil Companies," June 19, 2007.

16 U.S. Energy Information Administration, *Annual Energy Outlook 2006*, "Energy Technologies on the Horizon," www.eia.doe.gov/oiaf/archive/aeo06/issues.html.

17 Energy Information Administration, "Analysis of Crude Oil Production in the Arctic National Wildlife Refuge," May 2008, www.eia.doe.gov/oiaf/servicerpt/anwr/pdf/sroiaf(2008)03.pdf.

18 Ibid.

19 Ibid.

20 Energy Information Office, Short Term Energy Outlook, Price Summary, February 9, 2009, www.eia.doe.gov/steo.

21 Congressional Research Service, "Arctic National Wildlife Refuge: New Directions in the 100th Congress," February 8, 2007.

22 Energy Information Administration, "Impacts of Increased Access to Oil and Natural Gas Resources in the Lower 48 Federal Outer Continental Shelf," 2007, www.eia.doe.gov/oiaf/aeo/otheranalysis/ongr.html.

23 "The Costly Compromises of Oil from Sand," *New York Times*, January 6, 2009, www.nytimes.com/2009/01/07/business/07oilsands.html.

24 Congressional Research Service, "North American Oil Sands: History of Development, Prospects for the Future," December 11, 2007.

25 Government Accountability Office, "Energy Markets: Increasing Globalization of Petroleum Products Markets, Tightening Refinery De-

312   Notes

mand and Supply Balance, and Other Trends Have Implications for U.S. Energy Supply, Prices and Volatility," December 2007.

26 Government Accountability Office, "Energy Markets: Factors That Influence Gasoline Prices," May 2007.

27 Gallup Poll, "Several Industries Take Big Image Hit This Year," www.gallup.com/poll/109468/Several-Industries-Take-Big-Image-Hit-Year.aspx.

28 Gallup Organization, "U.S. Congress, Gouging Blamed Equally for Gas Prices," August 6, 2008, www.gallup.com/poll/109303/US-Congress-Gouging-Blamed-Equally-Gas-Prices.aspx.

29 Congressional Research Service, "Oil Industry Profit Review 2007," April 2008.

30 National Petroleum Council, "Geopolitics: Facing the Hard Truths about Energy," July 2007.

31 Congressional Research Service, "The Role of National Oil Companies in the International Oil Market," August 21, 2007.

32 National Energy Technology Laboratory, U.S. Department of Energy, "Peaking of World Oil Production: Impacts, Mitigation & Risk Management," February 2005, www.netl.doe.gov/publications/others/pdf/oil_peaking_netl.pdf.

33 Government Accountability Office, "Crude Oil: Uncertainty about Future Oil Supply Makes It Important to Develop a Strategy for Addressing a Peak and Decline in Oil Production," GAO-07-283 February 28, 2007.www.gao.gov/new.items/d07283.pdf.

34 Daniel Yergin, *The Prize: The Epic Quest for Oil, Money and Power* (New York: Free Press, 1991), p. 194.

35 Vaclav Smil, *Oil: A Beginner's Guide* (Oxford: Oneworld Publications, 2008), p. 166.

36 National Petroleum Council, "Facing the Hard Truths about Energy," July 2007, www.npchardtruthsreport.org.

37 Public Agenda, "Energy Learning Curve," April 2009, www.publicagenda.org/pages/energy-learning-curve.

38 John Gillespie Magee Jr., "High Flight."

39 See. for example, "Did Speculation Fuel Oil Price Swings?" *60 Minutes*, January 11, 2009, www.cbsnews.com/stories/2009/01/08/60minutes/main4707770.shtml.

40 Peter Pae, "Southwest Airlines Reaps Benefits of Fuel Hedging Strategy," *Los Angeles Times*, May 30, 2008.

41 David Koenig, "Analysts' Pessimism Hits Southwest Airlines Shares," *USA Today*, January 23, 2009, www.usatoday.com/travel/flights/2009-01-23-southwest-share-price_N.htm.

42 Interagency Task Force on Commodities Markets, "Interim Report on Crude Oil," June 2008, www.cftc.gov/stellent/groups/public/@newsroom/documents/file/itfinterimreportoncrudeoil0708.pdf.

43 Ibid.

44 "A Few Speculators Dominate Vast Market for Oil Trading," *Washington Post*, August 21, 2008, www.washingtonpost.com/wp-dyn/content/article/2008/08/20/AR2008082003898.html.

45 Richard A. Oppel Jr., "Word for Word/Energy Hogs; Enron Traders on Grandma Millie and Making Out Like Bandits," *New York Times*, June 13, 2004, http://query.nytimes.com/gst/fullpage.html?res=9E05E 6DE1330F930A25755C0A9629C8B63&scp=1&sq=enron,%20grand mothers&st=cse.

46 Federal Energy Regulatory Commission, "Staff Report on Price Manipulation in Western Markets," March 2003, www.ferc.gov/industries/ electric/indus-act/wec/enron/summary-findings.pdf.

47 Oppel, "Word for Word/Energy Hogs."

48 See, for example, David Kreutzer, "Oil Speculators Help Consumers at the Gas Pump," Heritage Foundation, WebMemo #2003, July 24, 2008, www.heritage.org/Research/EnergyandEnvironment/wm2003.cfm.

CHAPTER 7. YOU LOAD SIXTEEN TONS AND WHAT DO YOU GET?

1 EIA, "Where We Get Coal," www.eia.doe.gov/kids/energyfacts/sources/ non-renewable/coal.html#WhereWeGetCoal.

2 Ibid.

3 Quoted in George Monbiot, "The Stakes Could Not Be Higher. Everything Hinges on Stopping Coal," *Guardian*, August 5, 2008, www .guardian.co.uk/commentisfree/2008/aug/05/kingsnorthclimatecamp .climatechange.

4 Andrew C. Revkin, "Years Later, Climatologist Renews His Call for Action," *New York Times*, June 23, 2008, www.nytimes.com/2008/06/ 23/science/earth/23climate.html.

5 EIA, "How Coal Is Used," www.eia.doe.gov/kids/energyfacts/sources/ non-renewable/coal.html#uses.

6 EIA, "Energy Timelines: Coal," www.eia.doe.gov/kids/history/timelines/ coal.html.

7 This EIA chart shows coal as the cheapest energy source: www.eia .doe.gov/oiaf/aeo/figure_1.html.

8 BP, *Statistical Review of World Energy*, 2008, "Coal Consumption," www.bp.com/sectiongenericarticle.do?categoryId=9023786&contentId= 7044482.

9 Energy Information Administration, U.S. Coal Exports, www.eia.doe .gov/cneaf/coal/quarterly/html/t7p01p1.html.

10 Bureau of Labor Statistics, *Number and Rate of Fatal Occupational Injuries, by Industry Sector*, www.bls.gov/iif/oshwc/cfoi/cfch0005.pdf.

11 Jesse J. Holland, "Environmental groups sue over last-minute EPA rule easing limits on mountaintop mining," Associated Press and the *Cleveland Plain Dealer*, December 22, 2008, www.cleveland.com/nation/ index.ssf/2008/12/environmental_groups_sue_over.html#more.

12 Natural Resources Defense Council, "Land Facts: There Is No Such Thing as Clean Coal," www.nrdc.org/globalWarming/coal/coalmining .pdf.

13 EPA, "Environmental Progress," www.epa.gov/earthday/history.htm.

14 GAO, Statement of Jim Wells, Director, Natural Resources and Envi-

ronment, before the House Subcommittee on Energy and Resources, Committee on Government Reform, "Meeting Energy Demand in the 21st Century: Many Challenges and Key Questions," March 16, 2005.

15 Public Agenda, "Energy Learning Curve," April 2009, www.public agenda.org/pages/energy-learning-curve.

16 Environmental Protection Agency, Mercury: Controlling Power Plant Emissions—Overview, www.epa.gov/mercury/control_emissions/index .htm. Accessed May 2, 2009.

17 Cornelia Dean, "Environmentalists Advance on Emissions," *New York Times*, February 24, 2009.

18 Energy Information Administration, Greenhouse Gases, Climate Change, and Energy, Figure 4. U.S. Primary Energy Consumption and Carbon Dioxide Emissions, 2006, www.eia.doe.gov/bookshelf/ brochures/greenhouse/Chapter1.htm.

19 American Coalition for Clean Coal Electricity, www.cleancoalusa .org/docs/about.

20 Quoted in www.cbsnews.com/stories/2008/07/17/tech/livinggreen/main 4270123.shtml.

21 Norway, Official Site in India, "Carbon Dioxide Capture and Storage in Norway," August 26, 2998, www.norwayemb.org.in/Climate+and+ Environment/Climate/Carbon+capture.htm; Norway Mission to the European Community, "Major Research Programme for CO_2-Capture," August 18, 2008, www.eu-norway.org/Climate+change/Major+research+ programme+for+CO2-capture.htm.

22 CleanTechnica.com, "Carbon Capture and Storage Goes Online in Germany," September 10, 2008, http://cleantechnica.com/2008/09/10/ carbon-capture-and-storage-goes-online-in-germany.

23 Kimberly Kindy, "New Life for 'Clean Coal' Project," *Washington Post*, March 6, 2009.

24 IPCC, "Special Report on Carbon Dioxide Capture and Storage," www .mnp.nl/ipcc/pages_media/SRCCS-final/IPCCSpecialReportonCar bondioxideCaptureandStorage.htm.

25 Office of Management and Budget, The U.S. Department of Energy 2010 Budget: fact sheet, www.whitehouse.gov/omb/assets/fy2010_fact sheets/fy10_energy.pdf.

26 Massachusetts Institute of Technology (MIT), *The Future of Coal: Options for a Carbon-Constrained World*, 2007.

27 See, for example, Jeff Goodell, "Coal's New Technology: Panacea or Risky Gamble," *Yale Environment 360*, July 14, 2008.

28 Environmental Protection Agency, *Final Report of the Advanced Coal Technology Work Group*, January 29, 2008, http://earth1.epa.gov/air/ caaac/coaltech/2008_01_final_report.pdf.

29 NRDC, "No Time Like the Present: NRDC's Response to MIT's Future of Coal Report," March 2007, www.nrdc.org/globalWarming/coal/mit.pdf.

30 http://tonto.eia.doe.gov/ask/electricity_faqs.asp#coal_plants.

31 MIT, *The Future of Coal*.

32 Nichola Groom, "AEP CEO Says Coal Plants Must Be Retrofitted,"

Reuters, March 5, 2009, www.reuters.com/article/GCA-GreenBusiness/idUSTRE52469M20090305.

33 U.S. Energy Information Administration, "International Energy Prices for Households," accessed March 15, 2009, www.eia.doe.gov/emeu/international/elecprih.html.

34 Edmund Andrews, "Lawmakers Push for Big Subsidies for Coal Process," *New York Times*, May 29, 2007.

35 Ibid.

36 Environmental Protection Agency, "Greenhouse Gas Impacts of Expanded Renewable and Alternative Fuels Use," EPA420-F-07-035, April 2007, www.epa.gov/OMS/renewablefuels/420f07035.htm.

37 Ibid.

38 Ibid.

39 Peter Altman of National Environmental Trust, quoted in Edmund L. Andrews, "Lawmakers Push for Big Subsidies for Coal Process," *New York Times*, May 29, 2007, www.nytimes.com/2007/05/29/business/29coal.html.

40 According to American Association for the Advancement of Science, "coal-to-liquid technology is a well established process that does not necessitate future research to successfully produce liquid transportation fuels" (www.aaas.org/spp/cstc/briefs/coaltoliquid).

41 "In the Field: Pilot Project Uses Innovative Process to Capture CO_2 from Flue Gas," *EPRI Journal*, Spring 2008, pp. 4–5, http://mydocs.epri.com/docs/CorporateDocuments/EPRI_Journal/2008-Spring/1016422.pdf.

42 Energy Information Administration, 2007 Annual Coal Report, Employment, www.eia.doe.gov/cneaf/coal/page/acr/table18.pdf. Accessed May 2, 2009.

43 *All Things Considered*, National Public Radio, December 4, 2008, www.npr.org/templates/story/story.php?storyId=97825453.

CHAPTER 8. IT'S ALL RIGHT NOW (IN FACT, IT'S A GAS)

1 Leno quote from www.cleanskies.org.

2 Department of Energy, "The History of Natural Gas," www.fossil.energy.gov/education/energylessons/gas/gas_history.html.

3 EIA, Energy Basics 101 Homepage, "U.S. Primary Energy Consumption by Source and Sector, 2007," www.eia.doe.gov/basics/energybasics101.html.

4 See www.naturalgas.org/overview/uses_industry.asp.

5 Energy Information Administration, "Frequently Asked Questions, What percentage of homes in the U.S. use natural gas for heating?" http://tonto.eia.doe.gov/ask/ng_faqs.asp#ng_home_use.

6 Energy Information Administration, "Natural Gas—A Fossil Fuel. How Does Natural Gas Impact the Environment?" www.eia.doe.gov/kids/energyfacts/sources/non-renewable/naturalgas.html#naturalgas formation, accessed April 4, 2009.

7 GAO, "Economic and Other Implications of Switching from Coal to

Natural Gas at the Capitol Power Plant and at Electricity-Generating Units Nationwide," May 1, 2008, GAO-08-601R, p. 18.

8 EIA, *Greenhouse Gases, Climate Change, and Energy*, May 2008, Figure 4: "U.S. Primary Energy Consumption and Carbon Dioxide Emissions, 2006," www.eia.doe.gov/bookshelf/brochures/greenhouse/Chapter1.htm.

9 Department of Energy, "Energy Efficiency and Renewable Energy, Alternative and Advanced Fuels," www.afdc.energy.gov/afdc/fuels/natural_gas_what_is.html.

10 Environmental Protection Agency, *2009 Draft U.S. Greenhouse Gas Inventory Report, Draft Inventory of U.S. Greenhouse Gas Emissions and Sinks: 1990–2007*, Table ES-2, "Recent Trends in U.S. Greenhouse Gas Emissions and Sinks," February 2009, www.epa.gov/climatechange/emissions/downloads09/07ES.pdf, accessed April 4, 2009.

11 American Public Transportation Association, *2008 Public Transportation Fact Book*, Table 30, "Bus Power Sources," p. 31.

12 Energy Information Administration, "World Proved Reserves of Oil and Natural Gas, Most Recent Estimates," March 3, 2009, www.eia.doe.gov/emeu/international/reserves.html, accessed April 4, 2009.

13 Federal Energy Regulatory Commission, "Industries, LNG—The Importance of LNG," accessed March 7, 2009, www.ferc.gov/industries/lng/gen-info/import.asp.

14 Energy Information Administration, "U.S. Natural Gas Imports by Country," http://tonto.eia.doe.gov/dnav/ng/ng_move_impc_s1_a.htm, accessed April 4, 2009.

15 EIA, "Frequently Asked Questions: What Is the Volume of Natural Gas Reserves in the U.S. and Worldwide? Is There Enough to Meet Future Needs?" updated February 13, 2009, http://tonto.eia.doe.gov/ask/ng_faqs.asp#ng_reserves.

16 BP, *Statistical Review of World Energy*, June 2008, "Natural Gas Proved Reserves," p. 22.

17 See, for example, Gary Schmitt, "Energy Security, National Security, and Natural Gas," American Enterprise Institute for Public Policy Research, April 2006; Owen Matthews and Anna Nemtsova, "The Medvedev Doctrine," *Newsweek*, December 1, 2008.

18 Energy Information Administration, *Annual Energy Outlook 2009*, "Oil and Natural Gas Projections," p. 2, www.eia.doe.gov/oiaf/aeo/pdf/trend_4.pdf, accessed April 4, 2009.

19 Jad Mouawad, "Estimate Places Natural Gas Reserves 35% Higher," *New York Times*, June 18, 2009.

20 CleanSkies.org, "Study Finds Existing U.S. Natural Gas Supply Extends into the 22nd Century," www.cleanskies.org, accessed April 4, 2009.

21 Chris Nelder, "Will Arctic Oil, Natural Gas, MIT, Paris, and Pickens Save the Day?" *Energy and Capital*, August 6, 2008, www.energyandcapital.com/articles/arctic-oil-natural+gas/740.

22 EPA, "Underground Injection Control Program: What Is Hydraulic Fracturing?" www.epa.gov/ogwdw/uic/wells_hydrofrac.html.

23 "Natural Gas and Technology, Advances in the Exploration and Production Sector," www.naturalgas.org/environment/technology.asp.

24 See, for example, comments by Tracy Carluccio of Delaware Riverkeeper and Kate Sinding of the Natural Resources Defense Council in New York, both quoted in Mark Clayton, "Controversial Path to Possible Glut of Natural Gas," *Christian Science Monitor*, September 17, 2008.

25 Environmental Protection Agency, "Evaluation of Impacts to Underground Sources of Drinking Water by Hydraulic Fracturing of Coalbed Methane Resevoirs," June 2004, www.epa.gov/safewater/uic/pdfs/cbmstudy_attach_uic_final_fact_sheet.pdf, accessed April 4, 2009.

26 See, for example, Abraham Lusgarten, "Does Natural Gas Drilling Endanger Water Supplies," *Business Week*, November 11, 2008, www.businessweek.com/magazine/content/08_47/b4109000334640.htm, accessed April 4, 2009.

27 Deborah Elcock, "Environmental Policy and Regulatory Constraints to Natural Gas Production," Environmental Assessment Division, Argonne National Laboratory, December 2004, pp. 5–6.

28 National Petroleum Council, "Hard Truths: Facing the Hard Truths About Energy," July 2007.

29 U.S. Senate, Committee on Environment and Public Works, "Fact of the Day: Friday, Green Bigotry and Natural Gas," October 14, 2005, http://epw.senate.gov/public/index.cfm?FuseAction=PressRoom.Facts&ContentRecord_id=85B89AE4-0365-4A7B-9278-016EA87A958F.

30 National Wildlife Federation, "Help Save America's Public Lands," www.nwf.org/publiclands; Natural Resources Defense Council, "Issues: Wildlands," www.nrdc.org/land/use/qwest.asp#public.

31 Quoted in Juliet Eilperin, "Bureau Proposes Opening Up Utah Wilderness to Drilling," *Washington Post*, October 31, 2008.

32 Stephen Bloch of the Southern Utah Wilderness Alliance quoted in Juliet Eilperin, "Bureau Proposes Opening Up Utah Wilderness to Drilling," *Washington Post*, October 31, 2008.

33 PollingReport.com, "Energy: Pew Research Center for the People and the Press," June 18–29, 2008 ("Would you favor or oppose allowing oil and gas drilling in the Arctic National Wildlife Refuge in Alaska?") 50 percent favor; 43 percent oppose; 7 percent not sure; and "CNN/Opinion Research Corporation Poll. August 29–31, 2008" ("How do you feel about increased drilling for oil and natural gas offshore in U.S. waters? Do you strongly favor, mildly favor, mildly oppose or strongly oppose increased offshore drilling?"), 52 percent strongly favor; 22 percent mildly favor; 11 percent mildly oppose; 13 percent strongly oppose; 1 percent unsure, www.pollingreport.com/energy.htm.

34 Gallup Organization, March 2008: "I'm going to read you a list of environmental problems. As I read each one, please tell me if you personally worry about this problem a great deal, a fair amount, only a little, or not at all. How much do you personally worry about" www.pollingreport.com/enviro.htm.

35 Joel Achenbach, "Traditional Energy's Modern Boom," *Washington Post*, August 15, 2008, www.washingtonpost.com/wp-dyn/content/article/2008/08/14/AR2008081403321.htm.

CHAPTER 9. TIME FOR THE NUCLEAR OPTION?

1 EIA, "U.S. Nuclear Reactors," www.eia.doe.gov/neic/infosheets/nuclear.html; "States with Commercial Nuclear Reactors," www.eia.doe.gov/cneaf/nuclear/page/at_a_glance/reactors/states.html, accessed March 15, 2009.

2 Nuclear Energy Institute, "Resources and Stats, U.S. Nuclear Power Plants," www.nei.org/resourcesandstats/nuclear_statistics/usnuclear powerplants.

3 EIA, "Nuclear Basics 101," www.eia.doe.gov/basics/nuclear_basics.html.

4 Public Agenda, "Energy Learning Curve," April 2009, www.public agenda.org/pages/energy-learning-curve.

5 Toni Johnson, "Challenges for Nuclear Power Expansion," Council on Foreign Relations, August 11, 2008.

6 Matthew Wald, "After 35-Year Lull, Nuclear Power May Be in Early Stages of a Revival," *New York Times*, October 24, 2008.

7 Nuclear Regulatory Commission, "Expected New Nuclear Power Plant Applications," accessed March 15, 2009, www.nrc.gov/reactors/new-licensing/new-licensing-files/expected-new-rx-applications.pdf.

8 Elisabeth Bumiller, "McCain Sets Goal of 45 New Nuclear Reactors by 2030," *New York Times*, June 19, 2008, www.nytimes.com/2008/06/19/us/politics/19nuke.html.

9 "The Big Positive: Nuclear Produces Huge Amounts of Power with Zero Greenhouse," *Miami Herald*, October 20, 2008.

10 "Life After Death," *Economist*, June 21, 2008.

11 See, for example, Congressional Budget Office. "Nuclear Power's Role in Generating Electricity," May 2008.

12 Nuclear Energy Institute, "Sources of Emission-Free Electricity," www.nei.org/resourcesandstats/documentlibrary/protectingtheenvironment/graphicsandcharts/infographicemissionfree.

13 See Nuclear Energy Agency, *Nuclear Energy Today*, Chapter 9: "Nuclear Energy and Sustainable Development," Figure 9.4, www.nea.fr/html/pub/nuclearenergytoday/net/nuclear_energy_today_ch9.pdf, and States News Service, Statement by the International Atomic Energy Agency, "Nuclear Power Worldwide: Status and Outlook," September 11, 2008.

14 Public Agenda, "Energy Learning Curve." April 2009, www.publica-genda.org/pages/energy-learning-curve.

15 Government Accountability Office, "Meeting Energy Demands in the 21st Century," Testimony of Jim Wells, Natural Resources and Environment Before House Subcommittee on Energy and Resources, March 16, 2005, p. 18.

16 See Congressional Budget Office, *Nuclear Power's Role in Generating*

Electricity, May 2008, Figure 1-2, "Actual and Projected Natural Gas Prices, 1994 to 2030," www.cbo.gov/ftpdocs/91xx/doc9133/Chapter1.4.1.shtml#1088477.

17 Nuclear Energy Institute, *Financing New Nuclear Power Plants*, January 2009, www.nei.org/filefolder/Financing_New_Nuclear_Plants_January_2009.pdf.

18 Johnson, "Challenges for Nuclear Power Expansion."

19 Congressional Budget Office, *Nuclear Power's Role in Generating Electricity*, Table 2-1: "Projected and Actual Construction Costs for Nuclear Power Plants."

20 MIT, *The Future of Nuclear Power*, 2003.

21 Congressional Budget Office, *Nuclear Power's Role in Generating Electricity*; Table 2-1: Projected and Actual Construction Costs for Nuclear Power Plants, p. 17.

22 MIT, "The Future of Nuclear Power, 2003," http://web.mit.edu/nuclearpower/pdf/nuclearpower-full.pdf, and MIT, "Update of the MIT 2003 Future of Nuclear Power, 2009," http://web.mit.edu/nuclearpower/pdf/nuclearpower-update2009.pdf.

23 Florida Power and Light, "Clean and Safe Nuclear: Turkey Point 6 & 7 New Nuclear Project," www.fpl.com/environment/nuclear/pdf/nuclear_factsheet.pdf and www.fpl.com/environment/nuclear/about_turkey_point.shtml.

24 "The Radioactive Debate: The Big Positive: Nuclear Produces Huge Amounts of Power with Zero Greenhouse Gas Emissions," *Miami Herald*, October 20, 2008.

25 Ibid.

26 Florida Power and Light, "Clean and Safe Nuclear," www.fpl.com/environment/nuclear/tp.shtml.

27 Florida Power and Light, "Our Commitment to the Environment," www.fpl.com/environment/commitment.shtml.

28 "The Radioactive Debate: The Big Positive."

29 Ibid.

30 Ibid.

31 Mark Holt, Congressional Research Service, "Civilian Nuclear Waste Disposal," updated October 7, 2008, www.fas.org/sgp/crs/misc/RL33461.pdf.

32 www.fsmitha.com/timeline.html.

33 U.S. Department of Energy, Office of Civilian Radioactive Waste Management, "What Are Spent Nuclear Fuel and High-Level Radioactive Waste?" www.ocrwm.doe.gov/factsheets/doeymp0338.shtml, accessed April 4, 2009.

34 Ibid.

35 H. Josef Hebert, "Report: Nuclear Waste Disposal Will Cost US $96B," Associated Press Financial Wire, August 5, 2008.

36 Office of Management and Budget, Budget of the U.S. Government, Fiscal Year 2010, Terminations, Reductions, and Savings, page 68, www.whitehouse.gov/omb/budget/fy2010/assets/trs.pdf.

37 MIT, "The Future of Nuclear Power, 2003," http://web.mit.edu/nuclear power/pdf/nuclearpower-full.pdf.
38 MIT, 2009 update of "The Future of Nuclear Power, 2003," http://web.mit.edu/nuclearpower/pdf/nuclearpower-update2009.pdf.
39 Department of Energy, statement of Steven Chu, Secretary of Energy before the Committee on the Budget, United States Senate, March 11, 2009, www.energy.gov/news2009/6972.htm.
40 See the International Atomic Energy Agency, "How Safe Is Nuclear Energy?" www.iaea.org/blog/Infolog/?page_id=23#a1; Nuclear Energy Agency, "Society and Nuclear Energy: Towards a Better Understanding, 2008," www.nea.fr/html/general/policypapers.html; and Nuclear Energy Institute, "Resources and Stats, World Statistics," www.nei.org/resourcesandstats/nuclear_statistics/worldstatistics/.
41 NRC, "Fact Sheet on Oversight of Nuclear Power Plants," accessed March 15, 2009, www.nrc.gov/reading-rm/doc-collections/fact-sheets/oversight.html.
42 Ibid.
43 "Our View on Nuclear Security: Asleep on the Job," USA Today, October 11, 2007.
44 Roy Zimmerman, "Opposing View: We Moved Swiftly," USA Today, October 11, 2007.
45 "Ohio Nuclear Engineer Convicted of Lying about Cracks in Reactor," Environmental News Service, NBC News, San Diego, August 26, 2008.
46 GAO, "Nuclear Regulatory Commission: Oversight of Nuclear Power Plant Safety Has Improved, but Refinements Are Needed." September, 2006.
47 National Highway Traffic Safety Administration, "Overview Traffic Safety Facts 2007" and "Traffic Safety Facts 1994," www-nrd.nhtsa.dot.gov/Pubs/FARS94.PDF; www.nhtsa.dot.gov/portal/nhtsa_static_file_downloader.jsp?file=/staticfiles/DOT/NHTSA/NCSA/Content/TSF/2007/810993.pdf.
48 The National Academies, "Spent Fuel Stored in Pools at Some U.S. Nuclear Power Plants Potentially at Risk from Terrorist Attacks; Prompt Measures Needed to Reduce Vulnerabilities," press release, April 6, 2005, www8.nationalacademies.org/onpinews/newsitem.aspx?RecordID=11263.
49 Nuclear Regulatory Commission, "Backgrounder on Dirty Bombs, May, 2007, " www.nrc.gov/reading-rm/doc-collections/fact-sheets/dirty-bombs-bg.html.
50 See, for example, MIT, 2009 update of "The Future of Nuclear Power, 2003," http://web.mit.edu/nuclearpower/pdf/nuclearpower-update2009.pdf; and Jonathan Medalia for the Congressional Research Service, "A Brief Review of Threats and Responses, September, 2004," www.fas.org/irp/crs/RL32595.pdf.
51 Ibid.
52 Government Accountability Office, "Plants Have Upgraded Security, but the Nuclear Regulatory Commission Needs to Improve Its Process

for Revising the Design Basis Threat, " April 4, 2006, www.gao.gov/highlights/d06555thigh.pdf.

53 Nuclear Regulatory Commission, "Fact Sheet on Safety and Security Improvements at Nuclear Plants," www.nrc.gov/reading-rm/doc-col lections/fact-sheets/safety-security.html, accessed April 4, 2009.

54 Union of Concerned Scientists, testimony of Edwin S. Lyman, Ph.D., to the Senate Subcommittee on Clean Air, Climate Change and Nuclear Safety, May 26, 2005, www.ucsusa.org/assets/documents/nuclear_power/lyman_testimony_5-26-05.pdf.

55 Natural Resources Defense Council, "Position Paper: Commercial Nuclear Energy."

56 Nuclear Regulatory Commission, "Backgrounder on Chernobyl Nuclear Power Plant Accident," updated February 20, 2007, www.nrc .gov/reading-rm/doc-collections/fact-sheets/chernobyl-bg.html.

57 Nuclear Energy Institute, "Fact Sheet: Chernobyl and Its Consequences," November 2008, www.nei.org/filefolder/Chernobyl_Accident _and_its_Consequences_NOV08.pdf.

58 Environmental Protection Agency, "Chernobyl Power Plant, Ukraine," http://epa.gov/radiation/rert/chernobyl.html.

59 Department of Energy, Office of Health, Safety and Security, "Chernobyl Health Effects Studies," updated January 22, 2007, www.hss .energy.gov/HealthSafety/ihs/hstudies/chern_hes.html.

60 Nuclear Regulatory Commission, "Backgrounder on Chernobyl Nuclear Power Plant Accident"; World Health Organization, "Health Effects of the Chernobyl Accident," April 2006, www.who.int/media centre/factsheets/fs303/en/index.html.

61 World Health Organization, "Health Effects of the Chernobyl Accident," April 2006, www.who.int/mediacentre/factsheets/fs303/en/index.html.

62 World Health Organization, "Health Effects of the Chernobyl Accident."

63 Nuclear Energy Institute, "Fact Sheet: Chernobyl and Its Consequences."

64 Nuclear Energy Institute, "Chernobyl and Its Consequences," fact sheet, November 2008, www.nei.org/filefolder/Chernobyl_Accident_and_its_Consequences_NOV08.pdf.

65 Ibid.

66 Nuclear Regulatory Commission, "Backgrounder on Chernobyl Nuclear Power Plant Accident," www.nrc.gov/reading-rm/doc-collections/fact-sheets/chernobyl-bg.html.

67 Nuclear Regulatory Commission, "Backgrounder on Chernobyl Nuclear Power Plant Accident."

68 Ibid.

69 Nuclear Regulatory Commission, "Fact Sheet on the Three Mile Island Accident," March 2009, www.nrc.gov/reading-rm/doc-collections/fact-sheets/3mile-isle.html.

70 Ibid.

71 Nuclear Energy Institute, "The TMI 2 Accident: Its Impact, Its Lessons," December, 2007. www.nei.org/resourcesandstats/document library/safetyandsecurity/factsheet/tmi2accidentimpactlessons.

72 Nuclear Regulatory Commission, "Fact Sheet on the Three Mile Island Accident."

73 Ibid.

74 Associated Press, "U.S. Judge Throws Out Claims Against Three Mile Island Plant," *New York Times*, June 8, 1996.

75 See, for example, research conducted by Steven Wing of the University of North Carolina in *Environmental Health Perspective*, February 24, 1997, www.ehponline.org/docs/1997/105-1/wingabs.html.

76 Harvey Wasserman, "People Died at Three Mile Island," *Huffington Post*, March 26, 2009, www.huffingtonpost.com/harvey-wasserman/people-died-at-three-mile_b_179588.html.

77 Nuclear Regulatory Commission, "Fact Sheet on the Three Mile Island Accident."

78 Ibid.

79 "Live Discussion: Nuclear Power Since Three Mile Island," March 30, 1999, www.washingtonpost.com/wp-srv/national/talk/archive/hamilton0330.htm.

CHAPTER 10. AS LONG AS THE WIND BLOWS AND THE SUN SHINES

1 White House.gov Blog, "Electric," March 9, 2009, www.whitehouse.gov/blog/09/03/19/Electric/.

2 Energy Information Administration, "How Much Renewable Energy Do We Use?" August 21, 2008.

3 Energy Information Administration, *Annual Energy Outlook 2008*.

4 Energy Efficiency and Renewable Energy Lab, U.S. Department of Energy, "20 Percent Wind Energy by 2030: Increasing Wind Energy's Contribution to U.S. Energy Supply," May 2008.

5 Congressional Research Service, *Wind Power in the United States: Technology, Economic and Policy Issues*, June 20, 2008.

6 U.S. Department of Energy, *Annual Report on U.S. Wind Power Installation, Cost and Performance Trends 2007*, May 2008.

7 National Research Council, "Environmental Impacts of Wind-Energy Projects," May 2007, www.nap.edu/catalog/11935.html.

8 Ibid. For more on the local controversy, see www.saveoursound.org for the opponents, or www.capewind.org for the supporters.

9 Congressional Research Service, *Wind Power in the United States*.

10 National Renewable Energy Laboratory, "United States Annual Wind Resource Potential," accessed Oct. 19, 2008, www.nrel.gov/gis/wind.html.

11 Office of Energy Efficiency and Renewable Energy, U.S. Department of Energy, "20 Percent Wind Energy by 2030: Increasing Wind Energy's Contribution to U.S. Energy Supply," May 2008.

12 See Recovery.gov and Edison Electric Institute, "Transforming America's Power Industry: The Investment Challenge 2010–2030," November 2008, www.eei.org/ourissues/finance/Documents/Transforming_Americas_Power_Industry.pdf.

13 Congressional Research Service, *Wind Power in the United States*.

14 Ibid.

15 National Renewable Energy Laboratory, *Renewable Energy Price-Stability Benefits in Utility Green Power Programs*, August 2008, http://apps3.eere.energy.gov/greenpower/resources/pdfs/43532.pdf.

16 Royal Academy of Engineering, *The Cost of Generating Electricity*, www.raeng.org.uk/news/publications/list/reports/Cost_of_Generating_Electricity.pdf.

17 Interstate Renewable Energy Council, *U.S. Solar Market Trends 2007*, August 2008, www.irecusa.org/fileadmin/user_upload/NationalOut reachPubs/IREC%20Solar%20Market%20Trends%20August%202008_2.pdf. This just counts installations connected to the power grid (more on that later). Nobody really knows how many stand-alone installations there are.

18 Energy Efficiency and Renewable Energy, U.S. Department of Energy, "Net Metering Policies," accessed May 25, 2009, http://apps3.eere.energy.gov/greenpower/markets/netmetering.shtml.

19 Texas State Energy Conservation Office, "Texas Solar Energy," www.seco.cpa.state.tx.us/re_solar.htm.

20 U.S. Department of Energy, "PV FAQs," accessed October 26, 2008 www1.eere.energy.gov/solar/pdfs/35097.pdf.

21 Lisa Abend, "In Spain, A Solar-Powered Cemetery," *Time*, November 26, 2008, www.time.com/time/world/article/0,8599,1862364,00.html.

22 Ben Arnoldy, "Are Some Solar Projects No Longer 'Green'?" *Christian Science Monitor*, September 25, 2008, http://features.csmonitor.com/environment/2008/09/25/are-some-solar-projects-no-longer-%E2%80%98green%E2%80%99.

23 Bureau of Land Management, U.S. Department of the Interior, "BLM to Continue Accepting Solar Energy Applications," press release, July 2, 2008, www.blm.gov/wo/st/en/info/newsroom/2008/July/NR_07_02_2008.html.

24 U.S. Department of Energy, Office of Energy Efficiency and Renewable Energy, "Get Your Energy from the Sun: A Consumer's Guide." www.nrel.gov/docs/fy04osti/35297.pdf

25 United States Department of Energy, Solar Energy Technologies Program, "Solar Energy Industry Forecast: Perspectives on U.S. Solar Market Trajectory," June 24, 2008.

26 Interstate Renewable Energy Council, *U.S. Solar Market Trends 2007*, August 2008, www.irecusa.org/fileadmin/user_upload/NationalOut reachPubs/IRECSolarMarketTrendsAugust2008_2.pdf, and U.S. Department of Energy, Office of Energy Efficiency and Renewable Energy, "Get Your Energy from the Sun: A Consumer's Guide." www.nrel.gov/docs/fy04osti/35297.pdf.

27 United States Department of Energy, Solar Energy Technologies Program, "Solar Energy Industry Forecast: Perspectives on U.S. Solar Market Trajectory," June 24, 2008.

28 Robb Mandelbaum, "MIT Conference Is Bullish on Solar Power," *IEEE Spectrum*, April 2007, http://spectrum.ieee.org/apr07/5036.

29 Lori A. Bird, Karlynn S. Cory, and Blair G. Swezey, *Renewable Energy Price-Stability Benefits in Utility Green Power Programs*, NREL, http://apps3.eere.energy.gov/greenpower/resources/pdfs/43532.pdf.

30 U.S. Department of Energy, Solar Energy Technologies Program, "Solar Energy Industry Forecast: Perspectives on U.S. Solar Market Trajectory," June 24, 2008, www1.eere.energy.gov/solar/solar_america/pdfs/solar_market_evolution.pdf.

31 Center for American Progress, "Developing State Solar Photovoltaic Markets," January 2008.

32 Pew Center on Global Climate Change, "Renewable Portfolio Standards," accessed June 8, 2009, www.pewclimate.org/what_s_being_done/in_the_states/rps.cfm.

33 The *Wall Street Journal*, "Winds Shift for Renewable Energy As Oil Price Sinks, Money Gets Tight," Oct. 20, 2008.

34 Bureau of Reclamation, U.S. Department of Interior, "Hoover Dam Frequently Asked Questions and Answers: The Dam," accessed May 25, 2009, www.usbr.gov/lc/hooverdam/faqs/damfaqs.html.

35 Bureau of Reclamation, U.S. Department of Interior, "Hoover Dam Frequently Asked Questions and Answers: Hydropower at Hoover Dam," accessed May 25, 2009, www.usbr.gov/lc/hooverdam/faqs/powerfaq.html.

36 Congressional Research Service, "Energy: Selected Facts and Numbers," May 1, 2008.

37 Ibid.

38 Idaho National Laboratory, "Feasibility Assessment of the Water Energy Resources of the United States for New Low Power and Small Hydro Classes of Hydroelectric Plants, " January 2006, http://hydropower.id.doe.gov/resourceassessment/pdfs/main_report_appendix_a_final.pdf.

39 Idaho National Laboratory, "Hydropower Facts," accessed Nov. 2, 2008, http://hydropower.id.doe.gov/hydrofacts/index.shtml.

40 American Rivers, "Citizen Guides: Hydropower Dam Reform," www.americanrivers.org/site/PageServer?pagename=AR7_Guide_Hydropower.

41 Congressional Research Service, *Dam Removal: Issues, Considerations and Controversies*, June 19, 2006, http://assets.opencrs.com/rpts/RL33480_20060619.pdf.

42 EIA, "Frequently Asked Questions: How Much Electricity Does a Typical Nuclear Plant Generate?" http://tonto.eia.doe.gov/ask/faq.asp.

43 Energy Information Administration, "U.S. Energy Consumption by Source 2007," Scientific Forms of Energy, www.eia.doe.gov/kids/energyfacts/science/formsofenergy.html.

44 Colin F. Williams et al., "Assessment of Moderate- and High-Temperature Geothermal Resources of the United States," U.S. Geological Survey Fact Sheet 2008-3082, http://pubs.usgs.gov/fs/2008/3082.

45 Massachusetts Institute of Technology, *The Future of Geothermal Energy: Impact of Enhanced Geothermal Systems (EGS) on the United*

States in the 21st Century, 2006, www1.eere.energy.gov/geothermal/pdfs/future_geo_energy.pdf.

46 "Deep in Bedrock, Clean Energy and Quake Fears," June 23, 2009, *New York Times*, www.nytimes.com/2009/06/24/business/energy-environment/24geotherm.html.

47 Rick Sergel, "Executive Remarks: An Unprecedented Opportunity," speech to the Energy Future Coalition Grid Working Group, November 21, 2008, www.nerc.com/fileUploads/File/News/Executive-Remarks .112108.pdf.

48 Electric Power Research Institute, "Technology Primer: The Plug-In Electric Vehicle," 2007, http://mydocs.epri.com/docs/public/PHEV-Primer.pdf.

49 North American Electric Reliability Corporation, "2008 Long-Term Reliability Assessment," October 2008, www.nerc.com/files/LTRA2008 .pdf.

50 Congressional Research Service, "Electric Transmission: Approaches for Energizing a Sagging Industry," February. 12, 2007, http://opencrs .com/document/RL33875/2007-10-03.

51 Energy Information Administration, "Electric Power Industry Overview," accessed Dec. 30, 2008, www.eia.doe.gov/cneaf/electricity/page/prim2/toc2.html.

52 Electric Power Research Institute, "The Green Grid: Energy Savings and Carbon Emissions Reductions Enabled by a Smart Grid," June 2008, http://my.epri.com.

53 North American Electric Reliability Corporation, "2008 Long-Term Reliability Assessment," October 2008, www.nerc.com/files/LTRA2008 .pdf.

54 Electric Power Research Institute, "The Green Grid."

55 Edison Electric Institute, "Transforming America's Power Industry: The Investment Challenge 2010–2030," November 2008, www.eei.org/ourissues/finance/Documents/Transforming_Americas_Power_Indus try.pdf.

56 Edison Electric Institute, "Meeting U.S. Transmission Needs," July 2005, www.eei.org/ourissues/ElectricityTransmission/Documents/meet ing_trans_needs.pdf.

57 Congressional Research Service, "Electric Transmission: Approaches for Energizing a Sagging Industry," February 12, 2007, http://opencrs .com/document/RL33875/2007-10-03.

CHAPTER 11. NO PLACE LIKE [AN ENERGY EFFICIENT] HOME

1 EIA, "Energy Consumption by Sector Overview," www.eia.doe.gov/emeu/aer/pdf/pages/sec2_2.pdf.

2 Energy Information Administration, "Emissions of Greenhouse Gases," report, December 3, 2008, www.eia.doe.gov/oiaf/1605/ggrpt/carbon.html#emissions.

3 Ibid.

4 Department of Energy, Buildings Energy Data Book, "Characteristics of a Typical Single-Family Home," Table 2.2.11, www.btscoredatabook .net/TableView.aspx?table=2.2.11.

5 Energy Information Administration, "Frequently Asked Questions: How Much of the World's Energy Does the United States Use?" http://tonto.eia.doe.gov/ask/generalenergy_faqs.asp#energy_use_US, accessed April 5, 2009.

6 Energy Star, features of ENERGY STAR Qualified Homes, www.energy star.gov/index.cfm?c=new_homes.nh_features, accessed May 23, 2009.

7 DOE, Energy Star, "Behind the Walls Is What Really Counts," www .energystar.gov/index.cfm?c=behind_the_walls.btw_landing.

8 Environmental Protection Agency, "EPA Announces Energy Star Homes Reach Nearly 17 percent Market Share for 2008," July 2, 2009.

9 Ibid.

10 Eden Laikin, "Energy Star Laws on the Rise," *Newsday,* August 2, 2008; Gary Dymski, "Home Work: Want Energy Efficiency? Build It In," *Newsday,* May 17, 2007. See also Town of Brookhaven, LIPA Lauds Brookhaven's New E-Star Homes Law, April 25, 2007, www.brookhaven .org/PressRoom/tabid/56/mid/970/newsid970/261/Default.aspx.

11 Dymski, "Home Work."

12 Public Agenda, "Energy Learning Curve," April 2009, www.public agenda.org/pages/energy-learning-curve.

13 Cara Loriz, "Mandating Energy Efficiency in New Homes," *Shelter Island Reporter,* May 7, 2009, www2.timesreview.com/SIR/stories/i-m -tb-energy-0430.

14 "'Thirty Percent Solution' Defeated in Minneapolis," *Energy Design Update* 28, 11 (November 2008).

15 Department of Energy, Office of Energy Efficiency and Renewable Energy, Net Zero Energy Buildings Database, http://zeb.buildinggreen .com.

16 Timothy R. Homan, "U.S. Economy: Home Sales Fall, Record Drop in Prices," Bloomberg.com, www.bloomberg.com/apps/news?pid=2060 1068&sid=aE4zf50ys1zo&refer=home.

17 New York Academy of Sciences, "Academy E-Briefing: The Next Green Challenge: Energy Use in Existing Buildings," May 23, 2008, www .nyas.org/ebrief/miniEB.asp?ebriefID=729.

18 Energy Star, "Frequently Asked Questions: What Are Typically the Most Cost-Effective Improvements That I Can Make to My Home to Save Energy?" www.energystar.gov.

19 ACEEE, "Summary of Energy Efficiency Tax Incentives in the Energy Policy Act of 2005" and the Department of Housing and Urban Development, "Secretaries Donovan and Chu Announce Partnership to Help Working Families Weatherize Their Homes," February 27, 2009, www .hud.gov/recovery/2009/02/27/comms/pr09-016.cfm?CFID=15758401& CFTOKEN=77259353.

20 U.S. Census Bureau, *American Housing Survey for the United States:*

2005, Table 1A-1: "Introductory Characteristics—All Housing Units," www.census.gov/prod/2006pubs/h150-05.pdf.

21 Energy Star Home Advisor, www.energystar.gov/index.cfm?fuseaction= home_energy_advisor.showGetInput.

22 Energy Star, *2007 Annual Report*, Table 2: "Energy Star: Key Program Indicators, 2000 and 2007," www.energystar.gov/ia/partners/publica tions/pubdocs/2007%20Annual%20Report%20-%20Final%20-11-10-08.pdf.

23 Ibid, Figure 17: More than 4000 Buildings Have Earned the Energy Star.

24 Fred A. Bernstein, "Are McMansions Going Out of Style?" *New York Times*, October 2, 2005.

25 Energy Information Administration, "Assumptions to the Annual Energy Outlook 2009: Adjusting for the Size of Housing Units," Report #DOE/EIA-0554, March 2009, www.eia.doe.gov/oiaf/aeo/assumption/ pdf/residential.pdf.

26 EIA, "International Energy Outlook," Report #DOE/EIA-0484, June 2008, www.eia.doe.gov/oiaf/ieo/world.htm.

27 Bernstein, "Are McMansions Going Out of Style?"

28 Stephen Gandel, "The Incredible Shrinking House," CNNMoney.com, May 14, 2008.

29 Ibid.

30 National Association of Home Builders, "Older Buyers Seeking to Downsize—But Not by Much," press release, February 20, 2008.

31 Karen Shih, "For Skinny Houses, A Chilly Reception," *Baltimore Sun*, July 17, 2008.

32 GAO, "Energy Efficiency: Long-Standing Problems with DOE's Program for Setting Efficiency Standards Continue to Result in Foregone Energy Savings," GAO-07-42, June 2007.

33 John M. Broder, "Obama Orders New Rules to Raise Energy Efficiency," *New York Times*, February 5, 2009, www.nytimes.com/2009/ 02/06/us/politics/06energy.html?_r=1.

34 Energy Information Administration, "Energy in Brief: How Is My Electricity Generated, Delivered and Priced?" http://tonto.eia.doe.gov/ energy_in_brief/electricity.cfm.

35 Pew Research Center, "Pew Social & Demographic Trends, Topline for Selected Questions," October 3–19, 2008, http://pewsocialtrends.org/ assets/pdf/community-satisfaction-topline.pdf.

36 Blaine Harden, "Anti-Sprawl Laws, Property Rights Collide in Oregon," *Washington Post*, February 28, 2005.

37 Quoted in Joel Kotkin and Ali Modarres, "The Greening of Suburbia," *Seattle Times*. December 9, 2007, http://seattletimes.nwsource.com/ html/opinion/2004059264_sundaycities09.html.

38 James Howard Kunstler, "The Long Emergency: What's Going to Happen as We Start Running Out of Cheap Gas to Guzzle?" *Rolling Stone*, March 24, 2005.

39 Quoted in Kotkin and Modarres, "The Greening of Suburbia."

40 Congressional Research Service, "Energy Efficiency and the Rebound Effect: Does Increasing Efficiency Decrease Demand?" July 30, 2001.

41 Government Accountability Office, "Vehicle Fuel Economy: Reforming Fuel Economy Standards Could Help Reduce Oil Consumption by Cars and Light Trucks, and Other Options Could Complement These Standards," August 2007, www.gao.gov/new.items/d07921.pdf.

42 John Dimitropoulos, "Energy Productivity Improvements and the Rebound Effect: An Overview of the State of Knowledge," *Energy Policy*, September 2007; Horace Herring and Robin Roy, "Technological Innovation, Energy Efficient Design and the Rebound Effect," *Energy Policy*, January 2007.

CHAPTER 12. DRIVEN TO DISTRACTION

1 Art Buchwald, "How Un-American Can You Get?" *Have I Ever Lied to You?* (New York: Putnam, 1968).

2 U.S. Bureau of Transportation Statistics, *Pocket Guide to Transportation 2009*, January 2009, Table 4-6, www.bts.gov/publications/pocket_guide_to_transportation/2009.

3 "Obama to Toughen Rules on Emissions and Mileage," *New York Times*, May 18, 2009, www.nytimes.com/2009/05/19/business/19emissions.html.

4 Environmental Protection Agency, "Light-Duty Automotive Technology and Fuel Economy Trends: 1975 Through 2008," September 2008, www.epa.gov/otaq/fetrends.htm.

5 Ibid., Table 3: "Sales Fractions of MY1975, MY1988 and MY2007, Light-Duty Vehicles by Vehicle Size and Type," p. 21.

6 "How Slumping Market for SUVs Is Hurting Detroit's Bottom Line," *Wall Street Journal*, May 13, 2005, http://online.wsj.com/article/SB111595059715032621.html?mod=home_page_one_us.

7 Fueleconomy.gov, "Fuel Economy Guide 2009," www.fueleconomy.gov/feg/FEG2009.pdf.

8 www.consumerreports.org/cro/cars/new-cars/news/2008/10/affordable-hybrids/hybrid-owner-costs/affordable-hybrids-hybrid-owner-costs.htm.

9 Internal Revenue Service, "Alternative Motor Vehicle Credit," October 3, 2008, www.irs.gov/newsroom/article/0,,id=157632,00.html.

10 Consumer Reports, "Which Affordable Hybrids Save You the Most Money?," October 2008, www.consumerreports.org/cro/cars/new-cars/news/2008/10/affordable-hybrids/overview/affordable-hybrids-ov.htm; Mark Soltheim, "Hybrids Worth the Price," Kiplinger.com, September 1, 2008, www.kiplinger.com/columns/car/archive/2008/car0904.html.

11 Electric Power Research Institute, "Technology Primer: The Plug-In Electric Vehicle," 2007, http://mydocs.epri.com/docs/public/PHEV-Primer.pdf.

12 Congressional Research Service, *Advanced Vehicle Technologies*.

13 Joe Nocera, "Costly Toys, or a New Era for Drivers?" *New York Times*, July 19, 2008, www.nytimes.com/2008/07/19/business/19nocera.html.

14 "Electric Cars and a Smarter Grid," *New York Times*, February 17, 2009, http://greeninc.blogs.nytimes.com/2009/02/17/electric-cars-and-a-smarter-grid.

15 Congressional Research Service, *Hydrogen and Fuel Cell Vehicle R&D: FreedomCAR and the President's Hydrogen Fuel Initiative*, March 20, 2008.

16 Energy Information Administration, "Alternatives to Traditional Transportation Fuels 2007," April 2009, www.eia.doe.gov/cneaf/alternate/page/atftables/attf_v1.pdf www.

17 "U.S. Drops Research Into Fuel Cells for Cars," *New York Times*, May 7, 2009, www.nytimes.com/2009/05/08/science/earth/08energy.html.

18 Energy Information Administration, "Annual Energy Outlook 2009 Early Release Overview," January 2009, www.eia.doe.gov/oiaf/aeo/index.html.

19 Congressional Research Service, *Fuel Ethanol: Background and Public Policy Issues*, April 28, 2008.

20 Ibid., p. 9.

21 Alternative Fuels and Advanced Vehicles Data Center, U.S. Department of Energy, "Fuel Properties Comparison Chart," accessed February 1, 2009, www.afdc.energy.gov/afdc/progs/fuel_compare.php; Congressional Research Service, "Agriculture-Based Renewable Energy Production," March 7, 2007, Table 2.

22 Jason Hill et al., "Environmental, Economic, and Energetic Costs and Benefits of Biodiesel and Ethanol Biofuels," *Proceedings of the National Academies of Science*, July 25, 2006, www.pnas.org/content/103/30/11206.abstract.

23 Congressional Research Service, *Agriculture-Based Renewable Energy Production*, March 7, 2007, p. 14.

24 Goldemberg, Jose, "The Ethanol Program in Brazil," *Environmental Research Letters*, Institute of Physics, Nov. 24, 2006, www.iop.org/EJ/article/1748-9326/1/1/014008/erl6_1_014008.html#erl230223bib7.

25 U.S. Department of Agriculture, *The Economic Feasibility Of Ethanol Production from Sugar in the United States*, July 2006, www.usda.gov/oce/reports/energy/EthanolSugarFeasibilityReport3.pdf.

26 Hill et al., "Environmental, Economic, and Energetic Costs and Benefits of Biodiesel and Ethanol Biofuels."

27 Environmental News Network, "U.S. Will Fail to Meet Biofuels Mandate," Jan. 5, 2009, www.enn.com/energy/article/39003.

28 Alternative Fuels and Advanced Vehicles Data Center, U.S. Department of Energy, "Natural Gas Vehicle Availability," www.afdc.energy.gov/afdc/vehicles/natural_gas_availability.html, accessed March 19, 2009.

29 Honda.com, "2009 Civic Sedan" and "2009 Civic GX," http://automobiles.honda.com/civic-sedan and http://automobiles.honda.com/civic-gx, accessed May 25, 2009.

30 Energy Information Administration, "Alternatives to Traditional Transportation Fuels 2006," www.eia.doe.gov/cneaf/alternate/page/atftables/afvtransfuel_II.html.

31 Alternative Fuels and Advanced Vehicles Data Center, U.S. Department of Energy, "Fuel Properties Comparison Chart," www.afdc.energy.gov/afdc/fuels/properties.html.

32 Kate Galbraith, "Pickens Plan Stirs Debate, and Qualms," *New York Times*, August 5, 2008, www.nytimes.com/2008/08/05/business/05pickens.html.

33 National Research Council, "Effectiveness and Impact of Corporate Average Fuel Economy Standards," 2002, www.nap.edu/catalog.php?record_id=10172.

34 National Highway Traffic Safety Administration, "Overview Traffic Safety Facts 2007" and "Traffic Safety Facts 1994," www-nrd.nhtsa .dot.gov/Pubs/FARS94.PDF; www.nhtsa.dot.gov/portal/nhtsa_static_file_downloader.jsp?file=/staticfiles/DOT/NHTSA/NCSA/Content/TSF/2007/810993.pdf.

35 National Research Council, *Effectiveness and Impact of Corporate Average Fuel Economy Standards*, 2002, www.nap.edu/catalog.php?record_id=10172, p. 74.

36 Congressional Research Service, *Automobile and Light Truck Fuel Economy: The CAFÉ Standards*, May 7, 2008, http://opencrs.com/document/RL33413.

37 National Highway Traffic Safety Administration, "An Analysis of Motor Vehicle Rollover Crashes and Injury Outcomes," March 2007, www-nrd.nhtsa.dot.gov/Pubs/810741.PDF.

38 National Highway Traffic Safety Administration, "Overview Traffic Safety Facts 2007," www.nhtsa.dot.gov/portal/nhtsa_static_file_downloader.jsp?file=/staticfiles/DOT/NHTSA/NCSA/Content/TSF/2007/810993.pdf.

39 "Haiti: Thousands Protest Food Prices," April 8, 2008, Associated Press, www.nytimes.com/2008/04/08/world/americas/08briefs-THOUSANDSPRO_BRF.html.

40 "U.N. Food Meeting Ends With a Call for 'Urgent' Action," June 6, 2008, *New York Times*, www.nytimes.com/2008/06/06/world/06food.html.

41 World Bank, World Development Report 2008, "Biofuels: The Promise and the Risks," http://go.worldbank.org/UK40ECPQ20.

42 Congressional Budget Office, "The Impact of Ethanol Use on Food Prices and Greenhouse-Gas Emissions," April 2009, www.cbo.gov/ftpdocs/100xx/doc10057/04-08-Ethanol.pdf.

43 Organization for Economic Co-Operation and Development, "Biofuels: Is the Cure Worse Than the Disease?" September 2007, http://media.ft.com/cms/fb8b5078-5fdb-11dc-b0fe-0000779fd2ac.pdf.

44 U.N. Food and Agricultural Organization, "Food Outlook: Global Market Analysis, November 2008," www.fao.org/docrep/011/ai474e/ai474e00.htm.

45 Michael Grunwald, "The Clean Energy Scam," *Time*, March 27, 2008, www.time.com/time/magazine/article/0,9171,1725975,00.html.

46 National Academies of Science, "Increase in Ethanol Production from Corn Could Significantly Impact Water Quality and Availability if New Practices and Techniques Are Not Employed," October 10, 2007, www8.nationalacademies.org/onpinews/newsitem.aspx?Record ID=12039.

47 Hill et al., "Environmental, Economic, and Energetic Costs and Benefits of Biodiesel and Ethanol Biofuels."

CHAPTER 13. LOOKING FOR MR. WIZARD

1 Smithsonian Institution, "Remembering Gallery: American Inventors and Inventions," www.150.si.edu/150trav/remember/amerinv.htm.

2 http://inventors.about.com/od/weirdmuseums/ig/Photo-Gallery—Famous-Toys/Nintendo-Game-Boy-Patent.htm.

3 Matthew Power, "Kite-Sailing Tankers," and Vijay V. Vaitheeswaran, "Stan Ovshinsky," in "The Low Carbon Catalogue," *New York Times Magazine*, April 20, 2008; Andrew C. Revkin, "Husk Power for India," *New York Times*, December 24, 2008, www.nytimes.com/2009/01/04/education/edlife/ideas-huskpower-t.html?ref=edlife.

4 John M. Broder and Matthew L. Wald, "Big Science Role Is Seen in Global Warming Cure," *New York Times*, February 12, 2009.

5 MIT, *The Future of Coal*, 2007.

6 Abby Ellin, "Sweat Equity," in "The Low Carbon Catalogue," *New York Times Magazine*, April 20, 2008.

7 "Rising Above the Gathering Storm: Energizing and Employing America for a Brighter Economic Future," Statement of Norman R. Augustine Before the Committee on Science, U.S. House of Representatives, October 20, 2005, www7.nationalacademies.org/ocga/testimony/gathering_storm_energizing_and_employing_america2.asp.

8 Ibid.

9 See, for example, Alan Tonelson, "The Labor Shortage Hoax," American Economic Alert, January 27, 2006, www.americaneconomicalert.org/view_art.asp?Prod_ID=2205.

10 Sharon Begley, "Special Report: The Education Race: Or Maybe Major in Comp Lit?" *Newsweek*, August 9, 2008.

11 Ibid.

12 U.S. Department of Education, *National Mathematics Advisory Panel Final Report: Foundations for Success*, 2008, www.ed.gov/about/bdscomm/list/mathpanel/report/final-report.pdf, p. xii.

13 National Science Board, *A National Action Plan for Addressing the Critical Needs of the U.S. Science, Technology, Engineering, and Mathematics Education System*, October 2007, www.nsf.gov/nsb/documents/2007/stem_action.pdf.

14 U.S. Citizenship and Immigration Services, "Questions and Answers: USCIS Announces Interim Rule on H-1B Visas," updated May 9, 2008, www.uscis.gov/portal/site/uscis/menuitem.5af9bb95919f35e66f614176

543f6d1a/?vgnextoid=0189c9b9d87c8110VgnVCM1000004718190a RCRD.

15 Testimony of William A. Wulf, Ph.D., President, National Academy of Engineering, Before Subcommittee on Immigration, Border Security, and Claims, U.S. House of Representatives, September 15, 2005, www7.nationalacademies.org/ocga/testimony/Importance_of_Foreign _Scientists_and_Engineers_to_US.asp.

16 Thomas Friedman, "Tax Cuts for Teachers," New York Times, January 11, 2008, www.nytimes.com/2009/01/11/opinion/11friedman.html.

17 See, for example, comments by Illinois senator Richard J. Durbin in Robert Pear with Laurie J. Flynn, "High Tech Titans Strike Out on Immigration Bill," New York Times, June 25, 2007.

18 GAO, "H-1B Visa Program: More Oversight by Labor Can Improve Compliance with Program Requirements," statement of Sigurd R. Nilsen, Director, Education, Workforce, and Income Security, June 22, 2006, www.gao.gov/new.items/d06901t.pdf.

19 The Space Foundation, Space Technology Hall of Fame, www.spacet echhalloffame.org/history.

20 Hearings of the House Subcommittee on Energy and the Environ- ment, Department of Energy Fiscal Year 2009 Research and Develop- ment Budget Proposal, March 5, 2008, http://frwebgate.access.gpo .gov/cgi-bin/getdoc.cgi?dbname=110_house_hearings&docid=f:40940 .wais.pdf.

21 Dave Itzkoff, "Arbitration Dates Set in 'Speed-the-Plow,'" New York Times, March 26, 2009.

22 Association of American Universities, "The Year After Sputnik: U.S. Response, 1957–1958," September 18, 2007.

23 Ibid.

24 Ibid.

25 Association of American Universities, "The Year After Sputnik."

26 Louis Uchitelle, "'Buy America' in Stimulus (but Good Luck with That)," New York Times, February 20, 2009, www.nytimes.com/2009/ 02/21/business/21buy.html?ref=economy.

27 Dr. Paula Doe, "Explosive Growth Reshuffles Top 10 Solar Ranking," Renewable Energy World.com, September 12, 2008, www.renewableen ergyworld.com/rea/news/article/2008/09/explosive-growth-reshuffles -top-10-solar-ranking-53559.

28 Public Agenda, "Energy Learning Curve," April 2009, www.publica genda.org/pages/energy-learning-curve.

29 Electric Power Research Institute, "The Green Grid: Energy Savings and Carbon Emissions Reductions Enabled by a Smart Grid," June 2008, http://my.epri.com.

30 Stacey Chase, "Recession-Resistant Careers," Boston Globe, June 1, 2008.

31 Elena Foshay, "The New Apollo Program Fact Sheet," November 24, 2008, http://apolloalliance.org/new-apollo-program/data-points-the- new-apollo-program-fact-sheet.

32 Anya Kamenetz, "Ten Best Green Jobs for the Next Decade," *Fast Company*, January 13, 2009, www.fastcompany.com/articles/2009/01/best-green-jobs.html.

33 Ibid.

34 Azadeh Ansari, "Energy, Economy Create Balancing Act for Obama," CNN.com, January 30, 2009, www.cnn.com/2009/TECH/science/01/30/obama.climate.change/index.html.

35 Robert Samuelson, "The Bias Against Oil and Gas, " *Washington Post*, May 4, 2009, www.washingtonpost.com/wp-dyn/content/article/2009/05/03/AR2009050301849.html.

CHAPTER 14. SITTING ON TOP OF THE WORLD

1 World Energy Council, "Deciding the Future: Energy Policy Scenarios to 2050," 2007.

2 Global Network on Energy for Sustainable Development, "Summary for Policy Makers," March 2004.

3 Nathan S. Lewis, "Powering the Planet," *Engineering and Science*, 2007, http://eands.caltech.edu/articles/LXX2/powering.pdf.

4 United Nations Framework Convention on Climate Change, "Kyoto Protocol," accessed February 15, 2009, http://unfccc.int/kyoto_protocol/items/2830.php.

5 Senate Resolution 98, "A Resolution Expressing the Sense of the Senate Regarding the Conditions for the United States Becoming a Signatory to any International Agreement on Greenhouse Gas Emissions Under the United Nations Framework Convention on Climate Change," July 25, 1997, http://thomas.loc.gov/cgi-bin/bdquery/z?d105:SE00098:@@@L&summ2=m&.fs.

6 United Nations Framework Convention on Climate Change, "Fact Sheet: Copenhagen," accessed February 15, 2009, http://unfccc.int/files/press/backgrounders/application/pdf/fact_sheet_copenhagen_cop_15_cmp_5.pdf.

7 William Chandler, "Breaking the Suicide Pact: U.S.-China Cooperation on Climate Change," Carnegie Endowment for International Peace, May 2008.

8 Congressional Research Service, "China's Greenhouse Gas Emissions and Mitigation Policies," September 10, 2008, http://opencrs.com/document/RL34659.

9 Pew Center on Global Climate Change, "Common Challenge, Collaborative Response: A Roadmap for U.S.-China Cooperation on Energy and Climate Change," February 2009, www.pewclimate.org/US-China.

10 William Chandler, "Breaking the Suicide Pact: U.S.-China Cooperation on Climate Change," Carnegie Endowment for International Peace, May 2008.

11 Daniel Yergin, *The Prize: The Epic Quest for Oil, Money and Power* (Free Press, 1992), p. 183.

12 Council on Foreign Relations, "National Security Consequences of U.S. Oil Dependency," October 12, 2006.

13 Andrew E. Kramer, "Putin Orders Restored Oil Flow to Czechs," *New York Times*, July 22, 2008, www.nytimes.com/2008/07/22/world/europe/22czech.html.

14 Jeffrey Mankoff, *Eurasian Energy Security*, Council on Foreign Relations Special Report, February 2009, www.cfr.org/content/publications/attachments/Eurasia_CSR43.pdf.

15 Council on Foreign Relations, "U.S. Sanctions Biting Iran," January 23, 2007, www.cfr.org/publication/12478.

16 Council on Foreign Relations, "Tehran's Oil Dysfunction," February 16, 2007, www.cfr.org/publication/12625.

17 Heritage Foundation, "The Proposed Iran–Pakistan–India Gas Pipeline: An Unacceptable Risk to Regional Security," March 30, 2008, www.heritage.org/Research/AsiaandthePacific/upload/bg_2139.pdf.

18 Gross domestic product is the total amount of goods and services produced in a year—in other words, the total size of the U.S. economy.

19 Congressional Research Service, *Greenhouse Gas Emissions Drivers: Population, Economic Development and Growth, and Energy Use*, April 24, 2007.

20 Energy Information Administration, "Emissions of Greenhouse Gases in the U.S. 2007," December 3, 2008, www.eia.doe.gov/oiaf/1605/ggrpt.

21 Ibid.

22 United Nations Human Development Report 2007–8, *Fighting Climate Change: Human Solidarity in a Divided World*, http://hdr.undp.org/en/media/HDR_20072008_EN_Complete.pdf, Table 1.1.

23 U.S. Energy Information Administration, "U.S. Carbon Dioxide Emissions from Energy Sources 2008 Flash Estimate, " May 2009, www.eia.doe.gov/oiaf/1605/flash/pdf/flash.pdf.

24 Organization for Economic Co-Operation and Development, "Comparison of Gross Domestic Product (GDP) for OECD Member Countries," accessed June 10, 2008, www.oecd.org/dataoecd/48/4/37867909.pdf.

CHAPTER 15. SO NOW WHAT? IDEAS FROM THE LEFT, RIGHT, AND CENTER

1 http://tonto.eia.doe.gov/energy_in_brief/renewable_energy.cfm.

2 http://tonto.eia.doe.gov/energy_in_brief/foreign_oil_dependence.cfm.

3 http://tonto.eia.doe.gov/energy_in_brief/greenhouse_gas.cfm.

4 http://tonto.eia.doe.gov/energy_in_brief/greenhouse_gas.cfm.

5 www.gao.gov/htext/d08556t.html.

6 Thomas Friedman, "Win, Win, Win, Win, Win . . . ," *New York Times*, December 28, 2008, www.nytimes.com/2008/12/28/opinion/28friedman.html?_r=2&th&emc=th.

7 Robert J. Samuelson, "Stimulus for the Long Haul," *Washington Post*, editorial, October 29, 2008.

8 See, for example, Charles Krauthammer, "The Net-Zero Gas Tax," *Weekly Standard*, January 9, 2009, www.weeklystandard.com/Content/Public/Articles/000/000/015/949rsrgi.asp.

9 See, for example, Friedman, "Win, Win, Win, Win, Win . . ." *New York Times*, December 28, 2008, www.nytimes.com/2008/12/28/opinion/28 friedman.html?_r=2&th&emc=th.; Krauthammer, "The Net-Zero Gas Tax."

10 Public Agenda, "Energy Learning Curve," April 2009, www.public agenda.org/pages/energy-learning-curve.

11 Ibid.

12 Ben Lieberman and Jack Spencer, "Making Domestic Energy Affordable: A Memo to President-elect Obama," Heritage Foundation, December 8, 2008.

13 George Will, "The Gas Prices We Deserve," *Washington Post*, June 5, 2008, www.washingtonpost.com/wp-dyn/content/article/2008/06/04/AR2008060403052.html.

14 Nicolas Loris and Ben Lieberman, "Key Questions for Ken Salazar, Nominee for Department of Interior," Heritage Foundation, January 15, 2009, www.heritage.org/Research/EnergyandEnvironment/wm2224.cfm.

15 Robert J. Samuelson, "The Bias Against Oil and Gas," *Washington Post*, May 4, 2009, www.washingtonpost.com/wp-dyn/content/article/2009/05/03/AR2009050301849.html.

16 See, for example, Sierra Club, "Coastal Protection—May 22, 2006, Offshore Drilling Cheat Sheet," press release, May 22, 2006, www.sierraclub.org/pressroom/releases/pr2006-02-23.asp.

17 EPA, "Cap and Trade: Acid Rain Results," www.epa.gov/airmarkets/cap-trade/docs/ctresults.pdf.

18 George F. Will, "Carbon's Power Brokers," *Washington Post*, June 1, 2008, www.washingtonpost.com/wp-dyn/content/article/2008/05/30/AR2008053002521.html?referrer=emailarticle.

19 Lieberman and Spencer, "Making Domestic Energy Affordable."

20 See, for example, Mark Scott, "Is Europe Leading or Losing on CO_2 Emissions," *Business Week*, August 4, 2008; Olesya Dmitracova, "ECX Expects to Double CO2 Trading Volume in 2009," Reuters, May 11, 2009, www.reuters.com/article/GCA-GreenBusiness/idUSTRE54A3LD2009 0511; Price Carbon Campaign, "Frequently Asked Questions: How Is a Carbon Tax Better than Cap-and-Trade?" www.pricecarbon.org.

21 Price Carbon Campaign, "Frequently Asked Questions: What Is a Carbon Tax, Who Pays? and How Much Will the Tax Be?" www.pricecarbon.org.

22 Price Carbon Campaign, "Frequently Asked Questions: What Does a 'Revenue Neutral' Carbon Tax Mean?" www.pricecarbon.org.

23 David Frum, "The World Needs Pricier Oil for its Own Good," National Post.com, October 25, 2008, http://network.nationalpost.com/np/blogs/fullcomment/archive/2008/10/25/david-frum-the-world-needs-pricier-oil-for-its-own-good.aspx.

24 "Carbon Tax: Lesser of Two Evils," *Investor's Business Daily*, January 12, 2009.

25　Congressional Budget Office, *Policy Options for Reducing CO₂ Emissions*, February 2008, www.cbo.gov/ftpdocs/89xx/doc8934/toc.htm.

26　See, for example, MIT, *The Future of Coal*, 2007.

27　See, for example, Walter Alarkon, "Not All Senators Warming to Obama Cap-and-Trade Emissions Proposal," The Hill.com, March 4, 2009, http://thehill.com/leading-the-news/not-all-senators-warming-to-obama-cap-and-trade-emissions-proposal-2009-03-04.html.

28　EIA, *Greenhouse Gases, Climate Change, and Energy*, Figure 4: "U.S. Primary Energy Consumption and Carbon Dioxide Emissions, 2006," www.eia.doe.gov/bookshelf/brochures/greenhouse/Chapter1.htm.

29　Sierra Club, "Move Beyond Coal," www.sierraclub.org/coal.

30　Adam Satariano and Jeanmarie Todd, "Coal Ban May Avert 'Point of No Return,' Climate Scientists Say," Bloomberg.com, December 14, 2007, www.bloomberg.com/apps/news?pid=20601170&refer=home&sid=aH_peBEsdmEg.

31　See, for example, Nichola Groom, "AEP CEO Says Coal Plants Must Be Retrofitted," Reuters, March 5, 2009, www.reuters.com/article/GCA-GreenBusiness/idUSTRE52469M20090305.

32　Elisabeth Bumiller, "McCain Sets Goal of 45 New Nuclear Reactors by 2030," *New York Times*, June 19, 2008, www.nytimes.com/2008/06/19/us/politics/19nuke.html.

33　According to the Nuclear Energy Institute, it takes about four years (www.nei.org/filefolder/licensing_timeline_chart_Jan2009.jpg); See also Jack Spencer, "Time to Fast-Track New Nuclear Reactors," Heritage Foundation, WebMemo #2062, September 15, 2008, quotes the NRC as saying it needs forty-two months; www.heritage.org/Research/energyandenvironment/wm2062.cfm.

34　Jack Spencer and Nicolas Loris, "Grading the Gang of 10's Nuclear Energy Proposal," Heritage Foundation, September 8, 2008, www.heritage.org/Research/energyandenvironment/wm2053.cfm.

35　Ryan H. Wiser and Galen Barbose, "Renewables Portfolio Standards in the United States: A Status Report with Data through 2007: Report Summary," Lawrence Berkeley National Laboratory, April 2008, http://eetd.lbl.gov/ea/ems/reports/lbnl-154e-ppt-revised.pdf.

36　http://greeninc.blogs.nytimes.com/2009/02/20/next-up-a-renewable-portfolio-standard/.

37　Ryan H. Wiser and Galen Barbose, Lawrence Berkeley National Laboratory, "Renewables Portfolio Standards in the United States: A Status Report with Data Through 2007," report summary, April 2008, http://eetd.lbl.gov/ea/ems/reports/lbnl-154e-ppt-revised.pdf.

38　Ibid.

39　Kate Galbraith and Matthew L. Wald, "The Energy Challenge: Energy Goals a Moving Target for States," *New York Times*, December 5, 2008, www.nytimes.com/2008/12/05/business/05power.html.

40　See, for example, Stephanie I. Cohen, "Congress Struggles over How to Make Utilities Go Green," Dow Jones MarketWatch, June 13, 2007.

41　Cohen, "Congress Struggles over How to Make Utilities Go Green."

42 Ibid.

43 The Pickens Plan, www.pickensplan.com/didyouknow.

44 EIA, "Ethanol—A Renewable Fuel," www.eia.doe.gov/kids/energy facts/sources/renewable/ethanol.html.

45 Daniel J. Weiss and Nat Gryll, "Flex-Fuel Bait and Switch," Center for American Progress, June 18, 2007, www.americanprogress.org/issues/2007/06/flexfuel.html.

46 Jason Hill, et al., "Environmental, Economic, and Energetic Costs and Benefits of Biodiesel and Ethanol Biofuels," *Proceedings of the National Academies of Science*, July 25, 2006, www.pnas.org/content/103/30/11206.abstract.

47 "Tesla Charged about Possible Government Loan," Environmental Leader, February 13, 2009, www.environmentalleader.com/2009/02/13/tesla-ceo-charged-over-possible-government-loan.

48 Matt Nauman, "Tesla CEO: Federal loan for electric-car plant could come soon," *San Jose Mercury News*, February 11, 2009, www.mercury news.com/personalfinance/ci_11681303.

49 David E. Rodgers, "Building Energy Codes as a Response to Climate Change," Office of Technology Assessment, April 30, 2008, www.ase.org/uploaded_files/policy/codes_briefing_4-30-08/rodgers.pdf.

50 Lowell Ungar, Director of Policy, Alliance for Saving Energy, "Congressional Briefing: Building Energy Codes Prevent Climate Change," April 30, 2008, www.ase.org/uploaded_files/policy/codes_briefing_4-30-08/ungar.pdf.

51 "Administration Announces Nearly $8 Billion in Weatherization Funding and Energy Efficiency Grants," March 12, 2009, www.energy.gov/news2009/7015.htm.

52 Department of Energy, "The Manhattan Project: An Interactive History, Establishing Los Alamos," www.cfo.doe.gov/me70/manhattan/establishing_los_alamos.htm.

53 MIT Energy Initiative, "About MITEI," http://web.mit.edu/mitei/about/index.html.

54 See, for example, James Pethokoukis, "Do We Need an Energy 'Manhattan Project'?" *U.S. News and World Report*, May 30, 2008, www.usnews.com/blogs/capital-commerce/2008/5/30/do-we-need-an-energy-manhattan-project.html; Dan Greenberg," A Manhattan or Apollo Project for Energy? What Nonsense," *Chronicle of Higher Education*, April 6, 2008, http://chronicle.com/review/brainstorm/greenberg/a-manhattan-or-apollo-project-for-energy-what-nonsense.

55 Congressional Budget Office, "The Budgetary Implications of NASA's Current Plans for Space Exploration," p. 16 (Figure 9), April 2009, www.cbo.gov/ftpdocs/100xx/doc10051/04-15-NASA.pdf.

56 Erik Sofge, "Rebuilding America Special Report: How to Fix U.S. Infrastructure," *Popular Mechanics*, May 2008, www.popularmechanics.com/technology/transportation/4258053.html?page=3&series=53.

57 Ibid.

58 Edison Electric Institute, "Transforming America's Power Industry: The Investment Challenge 2010–2030," November 2008, www.eei.org/ourissues/finance/Documents/Transforming_Americas_Power_Industry.pdf.

59 Glenn Croston, "Stimulus Package Has Green for Clean Energy," MSNBC.com, March 13, 2009, www.msnbc.msn.com/id/29678865.

60 Matthew L. Wald, "Hurdles (Not Financial Ones) Await Electric Grid Update," *New York Times*, February 7, 2009.

61 Juliet Ielperin and Lyndsey Layton, "McCain Has Plan to Make Government More Green," *Washington Post*, June 25, 2008.

62 Ibid.

63 U.S. General Services Administration, GSA since 1949, www.gsa.gov/Portal/gsa/ep/contentView.do?programId=16117&channelId=-24830&ooid=10482&contentId=23723&pageTypeId=8199&contentType=GSA_BASIC&programPage=%2Fep%2Fprogram%2FgsaBasic.jsp&P=.

64 Monday Business Briefing, "President-elect Barack Obama's Energy and Environmental Policies, December 1, 2008.

65 Kimberly Kindy and Dan Keating, "Problems Plague U.S. Flex-fuel Fleet," *Washington Post*, November 23, 2008.

66 Ibid.

67 Ibid.

68 Public Agenda, "Energy Learning Curve."

69 Bureau of Transportation Statistics, *Pocket Guide to Transportation Statistics*, Chart 4-3: "Passenger Miles: 1990–2006," page 19.

70 American Public Transportation Association, "Public Transportation Facts," www.apta.com/media/facts.cfm#hw02.

71 Mathew L. Wald, "Amtrak's Own Board Sows Alarm About System's Future," *New York Times*, February 20, 2005.

72 Government Accountability Office, "Inventory of Major Federal Energy Programs and Status of Policy Recommendations," June 2005, www.gao.gov/new.items/d05379.pdf.

73 Dan Kish, "A Conservative Energy Policy," *Human Events*, June 5, 2008, www.humanevents.com/article.php?id=26833.

74 Ibid.

75 Michael Gerson, "Hope on Climate Change? Here's Why," *Washington Post*, August 15, 2007.

76 Ibid.

77 Reuters, "Pickens Predicts Oil to Top $100 by End of 2010," January 7, 2009, www.cnbc.com/id/28534827.

78 Submitted in Testimony of Michael Goo, Climate Legislative Director, Natural Resources Defense Council, Before the Subcommittee on Energy and Air Quality, House Committee on Energy and Commerce, June 19, 2008, http://docs.nrdc.org/globalWarming/files/glo_08061901a.pdf.

79 "Obama's Speech on the Economy," *New York Times*, January 8, 2009, www.nytimes.com/2009/01/08/us/politics/08text-obama.html?_r=1&ref=politics.

80 Ibid.

81 See www.recovery.gov/?q=content/act.

82 Barack Obama, acceptance speech at the Democratic National Convention, August 28, 2008, www.demconvention.com/barack-obama.

83 Barack Obama, "Real Leadership for a Clean Energy Future," October 8, 2007, posted at http://gristmill.grist.org/story/2007/10/8/122051/1952007.

84 Obama, acceptance speech at the Democratic National Convention.

85 Obama, "Real Leadership for a Clean Energy Future."

86 Ibid.

87 Ibid.

88 Ibid.

CHAPTER 16: THE REALITY SHOW

1 IPCC, *Climate Change 2007: The Physical Science Basis. Contribution of Working Group I to the Fourth Assessment Report of the Intergovernmental Panel on Climate Change* (Cambridge: Cambridge University Press, 2007); "Frequently Asked Question 10.3: If Emissions of Greenhouse Gases Are Reduced, How Quickly Do Their Concentrations in the Atmosphere Decrease?" www.ipcc.ch/pdf/assessment-report/ar4/wg1/ar4-wg1-faqs.pdf.

2 Energy Information Administration, *Greenhouse Gases, Climate Change, and Energy,* Brochure #DOE/EIA-X012, May 2008, www.eia.doe.gov/bookshelf/brochures/greenhouse/Chapter1.htm.

3 See, for example, David Leonhardt, "The Big Fix," *New York Times Magazine,* January 27, 2009, www.nytimes.com/2009/02/01/magazine/01Economy-t.html?_r=1.

4 John M. Broder, "Geography Is Dividing Democrats over Energy," *New York Times,* January 26, 2009, www.nytimes.com/2009/01/27/science/earth/27coal.html.

5 Ibid.

6 www.imdb.com/title/tt0070723/quotes.

7 www.imdb.com/title/tt0112384/quotes.

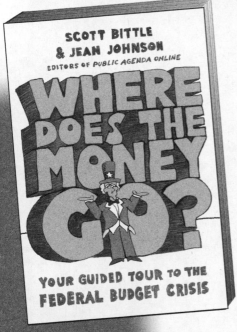